PRACTICAL COURSE IN ZOOLOGY

For
F.Y.B.Sc. ZOOLOGY, Semester - I and II
Course ZO-113 and ZO-123
New Syllabus as per CBCS Pattern, Credit-1.5-1.5, June 2019

Dr. Kishore R. Pawar
M.Sc., Ph.D.
Former Principal,
Karmveer Abasaheb Alias
N. M. Sonawane Arts, Science
and Commerce College,
Satana, Dist. Nashik.

Dr. Ashok E. Desai
M.Sc., Ph.D.
Former Associate Professor,
P. G. Department of Zoology,
KTHM College,
Nashik - 422002.

N5017

Practical Zoology Animal Diversity I and Ecology | **ISBN 978-93-89533-01-9**

First Edition : September 2019

© : Authors

Published By:
NIRALI PRAKASHAN
Abhyudaya Pragati, 1312, Shivaji Nagar
Off J.M. Road, PUNE – 411005
Tel - (020) 25512336/37/39, Fax - (020) 25511379
Email : niralipune@pragationline.com

DISTRIBUTION CENTRES

PUNE

Nirali Prakashan : 119, Budhwar Peth, Jogeshwari Mandir Lane, Pune 411002,
(For orders within Pune) Maharashtra, Tel : (020) 2445 2044, Mobile : 9657703145
Email : niralilocal@pragationline.com

Nirali Prakashan : S. No. 28/27, Dhayari, Near Asian College Pune 411041
(For orders outside Pune) Tel : (020) 24690204; Mobile : 9657703143
Email : bookorder@pragationline.com

MUMBAI

Nirali Prakashan : 385, S.V.P. Road, Rasdhara Co-op. Hsg. Society Ltd.,
Girgaum, Mumbai 400004, Maharashtra;
Mobile : 9320129587 Tel : (022) 2385 6339 / 2386 9976,
Fax : (022) 2386 9976
Email : niralimumbai@pragationline.com

DISTRIBUTION BRANCHES

JALGAON

Nirali Prakashan : 34, V. V. Golani Market, Navi Peth, Jalgaon 425001,
Maharashtra, Tel : (0257) 222 0395, Mob : 94234 91860;
Email : niralijalgaon@pragationline.com

KOLHAPUR

Nirali Prakashan : New Mahadvar Road, Kedar Plaza, 1st Floor Opp. IDBI Bank,
Kolhapur 416 012, Maharashtra. Mob : 9850046155;
Email : niralikolhapur@pragationline.com

NAGPUR

Nirali Prakashan : Above Maratha Mandir, Shop No. 3, First Floor,
Rani Jhanshi Square, Sitabuldi, Nagpur 440012, Maharashtra
Tel : (0712) 254 7129;
Email : niralinagpur@pragationline.com

DELHI

Nirali Prakashan : 4593/15, Basement, Agarwal Lane, Ansari Road, Daryaganj
Near Times of India Building, New Delhi 110002
Mob : 08505972553, Email : niralidelhi@pragationline.com

BENGALURU

Nirali Prakashan : Maitri Ground Floor, Jaya Apartments, No. 99, 6th Cross,
6th Main, Malleswaram, Bengaluru 560003, Karnataka;
Mob : 9449043034
Email: niralibangalore@pragationline.com

Other Branches : Hyderabad, Chennai

niralipune@pragationline.com | www.pragationline.com
Also find us on www.facebook.com/niralibooks

This book is dedicated

to our parents

Dr. Kishore R. Pawar
Dr. Ashok E. Desai

Preface ...

It gives us great pleasure to present this book **"Practical Zoology"** for the students of First Year B.Sc. Zoology, Savitribai Phule Pune University. This book is written according to the new CBCS syllabus of June 2019.

We have tried our best to present the subject matter in an easy style and in a comprehensive manner. The subject matter is profusely illustrated with a number of clear and labelled diagrams. We sincerely feel that this book will fulfill the requirements of the students as well as teachers. While preparing this book several standard reference books and text books have been consulted. Emphasis has been laid on furnishing maximum information required for students in a simple and lucid language. Zoology is an interesting subject because the animal world is full of diversity, adaptations, habits and habitats and behaviour. There are several textbooks and reference books available written by Indian and foreign authors but these books are too costly and majority of the students are unable to purchase them for comprehensive study.

We express our sincere thanks to Shri. Dineshbhai Furia, Shri. Jignesh Furia, Mr. Malik Shaikh, Mrs. Anjali Muley, Mrs. Roshan Shaikh and the entire staff of Nirali Prakashan for taking keen interest in the publication of this book and bringing out the book on time.

We shall gratefully accept constructive suggestions from the teachers as well as students for improvement of this book.

Prin. Dr. Kishore R. Pawar

Dr. Ashok E. Desai

Syllabus ...

Animal Diversity - I (Semester - I) (1.5 Credits - 45 Hours)

1. Museum Study of phylum Protozoa: *Euglena, Paramecium, Amoeba, Plasmodium sp.*
2. Museum study of Phylum Porifera: *Sycon, Euplectella, Chalina,* Spongilla.
3. Museum study of phylum Cnidaria: *Hydra, Physalia, Aurelio, Metridium.*
4. Museum study of phylum Platyhelminthes: *Planeria, Faciola hepatica, Taenia solium.*
5. Study of Paramecium: Culture, External morphology, Conjugation and Binary fission.
6. Study of permanent slides: Spicules and Gemmules in Sponges, T.S. of *Sycon,* T.S. of Hydra, *Taenia solium*: Scolex, Gravid proglottid.
7. Identification of any three museum specimen with help of taxonomic identification key.
8. Visit to Zoological survey of India/ Museum/National Park.

Animal Ecology

1. Estimation of Dissolved oxygen from given water sample.
2. Estimation of Water Alkalinity from given water sample.
3. Study of animal community structure by quadrate method (Field or Simulation).
4. Determination of density, frequency and abundance of species by quadrat method.
5. Study of microscopic fauna of freshwater ecosystem (from pond).
6. Estimation of water holding capacity of given soil sample.
7. Estimation of dissolved and free carbon dioxide from water sample.
8. Study of Eutrophication in lake/river.

Animal Diversity - II (Semester - II) (1.5 credits - 45 Hours)

1. Museum study of Phylum Aschelminthes: *Ascaris lumbricoides.*
2. Museum study of phylum Annelida: *Neries,* Earthworm, Leech.
3. Museum study of phylum Arthropods: Prawn, Cockroach, Centipede, Millipede, Crab.
4. Museum study of phylum Molluscs: *Pila, Chiton,* Bivalve, Octopus.
5. Museum study of phylum Echinodermata: Sea Star, Sea urchin, Brittle Star, Sea cucumber.
6. Study of permanent slides: Mouthparts of Insects - Mandibulate, Piercing and sucking, Chewing and Lapping.
7. Types of Shells in Molluscs. *Pila,* Bivalve, Chiton, Sepia.
8. Economic importance of honey bees, Lac insects silk worms, red cotton bug, Anopheles mosquito.
9. Earthworm: Vermicomposting bin preparation and maintenance.
10. Visit to a vermicomposting unit/ field for insect pest collection and its identification.

Cell Biology

1. Study of Microscope: Simple and Compound.
2. Micrometry: Measurement of microscopic objects.
3. Study of cell: Preparation of temporary mount of human buccal epithelial cells.
4. Preparation of blood smears to observe the blood cells.
5. Temporary preparation of mitotic cell from onion roots
6. Study of Cell Organelles (any three) by using microphotographs.

***

Contents ...

✳✳✳

SECTION - I : ANIMAL DIVERSITY - I
(Semester - I)

Practical **1** ...

Aim :

To study the classification with reasons of the following :

Phylum - Protozoa : *Euglena, Paramecium, Amoeba and Plasmodium sp.*

Thousands of different kinds of animals have been identified and described. And the number is continuously increasing. Animals are present with a considerable *diversity* in shape, structure, habit, habitat and mode of life. The variations in animals is the problem to scientist for their study. Taxonomy aims at describing all the known forms of animals, providing names (nomenclature) to them and establishing their relationships, so that they may be classified easily.

Five Kingdom Classification (Robert Whittaker, 1969) :

Robert Whittaker in 1969, presented alternative, basic classification for organisms which reflects phylogeny accurately and clearly; and is being widely used now.

The *Five kingdom classification* is based on three criteria :

(i)   Complexity of cell (Prokaryotic or Eukaryotic).

(ii)   Complexity of the body of the organisms (unicellular or multicellular).

(iii)   Mode of nutrition (autotrophic or heterotrophic).

According to this new scheme of classification, the living organisms are divided into two *super kingdoms – Prokaryote* and *Eukaryote.*

Prokaryote involves only one kingdom – **Monera** and Eukaryote is divided into four kingdoms; (i) **Protista,** (ii) **Plantae,** (iii) **Fungi** and (iv) **Animalia.**

The five kingdom classification has no place for viruses. This is the drawback of this method of classification. The kingdom Animalia comprises 33 phylums/sub-kingdoms.

PHYLUM : PROTOZOA

General Characters :

1.   Protozoan are small *unicellular* and microsocpic, exhibit a great variety of shape.

2.   They may be *solitary* or *colonial* in nature.

3.   Body either *naked* or covered by *pellicle*/shells with internal skeleton.

4.   The cytoplasm is differentiated into an inner *endoplasm* and outer *ecotplasm.*

5.   Locomotory organelles are psuedopodia (in Rhizopoda), *Flagella* (Flagellata) or *Cilia* (Ciliata).

6.　Nutrition may be *holophytic* (plant-like), holozoic (animal like), saprozoic or parasitic.

7.　Protozoan do not have specific respiratory and excretory organs. Both carried thorugh general body surface by *diffusion*.

8.　Asexual reproduction by binnary fission or budding and sexual reproduction by means of copulation or syngamy and conjugation.

9.　Life history is complicated in some cases with alternation of asexual and sexual phases.

10.　Protozoa exhibit two modes of life, free living or parasitic.

Classification of Phylum – *Protozoa*; (Gr; Protos = First; Zoon = Animal)

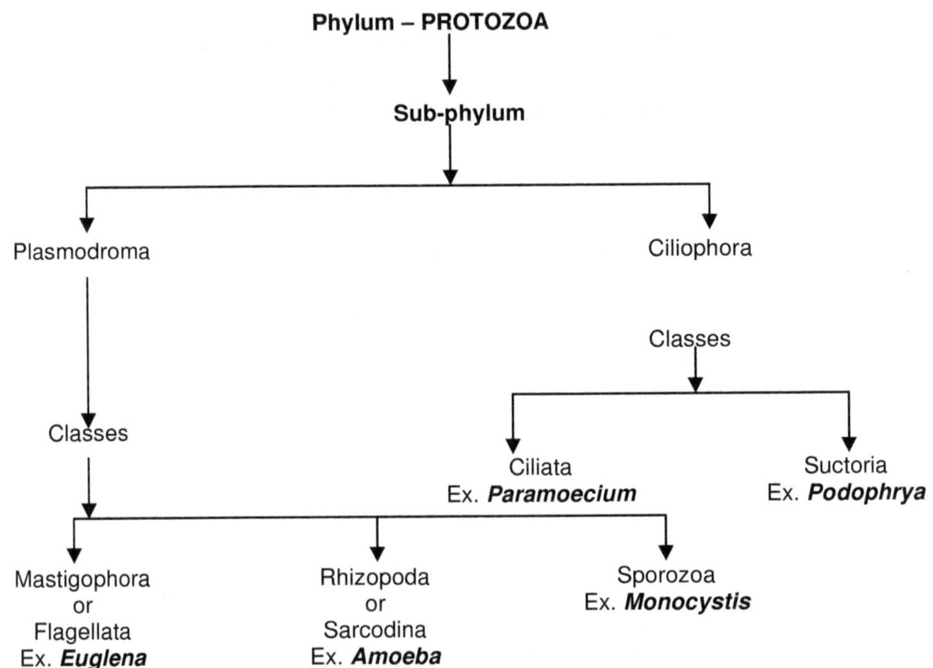

1. *EUGLENA*

Classification :

Phylum	:	*Protozoa* – Microscopic and unicellular.
Sub-phylum	:	*Plasmodroma* – Locomotory organelle, pseudopodia or flagella.
Class	:	*Mastigophora* – Locomotion by flagella.
Order	:	*Euglenoidina* – Body coverd by pellicle, cytostome and cytopharynx are present; Flagella one or two; chromatophores are present.
Genus	:	*Euglena*

Habit and Habitat :

Euglena is a common, solitary and free living flagellate found in freshwater ponds, ditches, lakes rich in nitrogeneous organic matters. They produce a green scum on the surface of ponds.

Distribution :

There are about 100 species of *Euglena* widely distributed (cosmopolitan) all over the world in polysaprobic fresh water. The most common *Euglena* is *Euglena viridis*.

Salient Features :

1. Body is covered by straited *pellicle*; which is flexible hence does *euglunoid movement.*

2. Anterior end is blund bears a funnel like *cytostome* and tubular *cytopharynx* while posterior end is pointed.

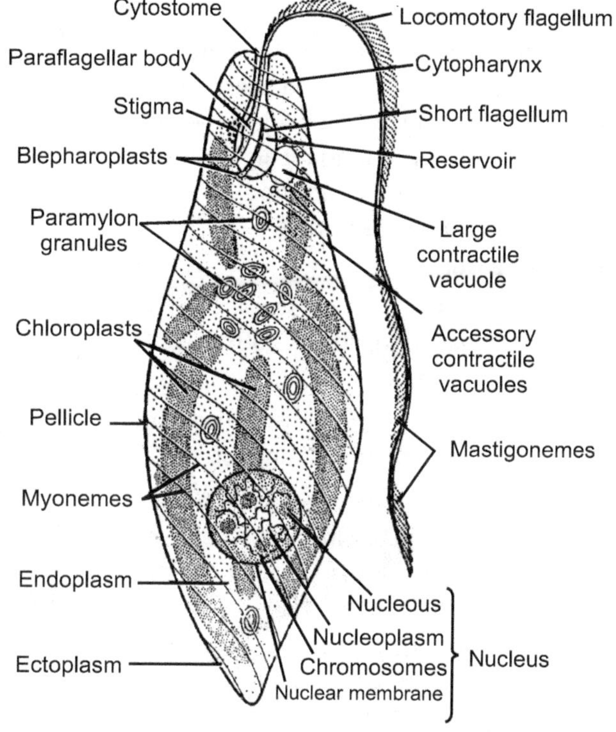

Fig. 1.1 : *Euglena*

3. Single whip-like *flagellum* arising from the base of the reservoir.
4. *Nucleus* is large located at the posterior region of the body.
5. The cytoplasm is differentiated into inner *endoplasm* and outer *ectoplasm*.
6. Nutrition is *holophytic* or *saprophytic*.
7. Reproduction takes place asexually by longitudinal binnary fission.
8. Body is spindle shaped and measures from 50 to 100 µ in length.

2. PARAMOECIUM

Classification :

Kingdom	:	Protista
Phylum	:	Ciliophora
Class	:	Ciliata
Subclass	:	Holotrichia
Order	:	Hymonostomatidia
Family	:	Paramecidae
Genus	:	*Paramoecium*
Species	:	*caudatum*

Habits and Habitat :

Paramoecium is cosmopolitan in distribution. It is found in fresh water, lakes, ponds, puddles, sewage pipes and rice fields. It is abundant in stagnant water containing decaying organic matter. It is omnivore in diet.

Salient Features :

1. It is a microscopic animal measuring about 200 to 300 µ in length.

2. It feeds on tiny protozoans, bacteria, bites of animal.

3. *Paramoecium* looks like the sole of slipper and that is why the animal is also called "Slipper animalcule".

4. It is streamlined, asymmetrical with well marked oral and aboral surfaces on the body.

5. Cytoplasm divides into ectoplasm and endoplasm. Ectoplasm secretes pellicle which gives definite shape to the body along with protection being gelatinous and elastic in nature.

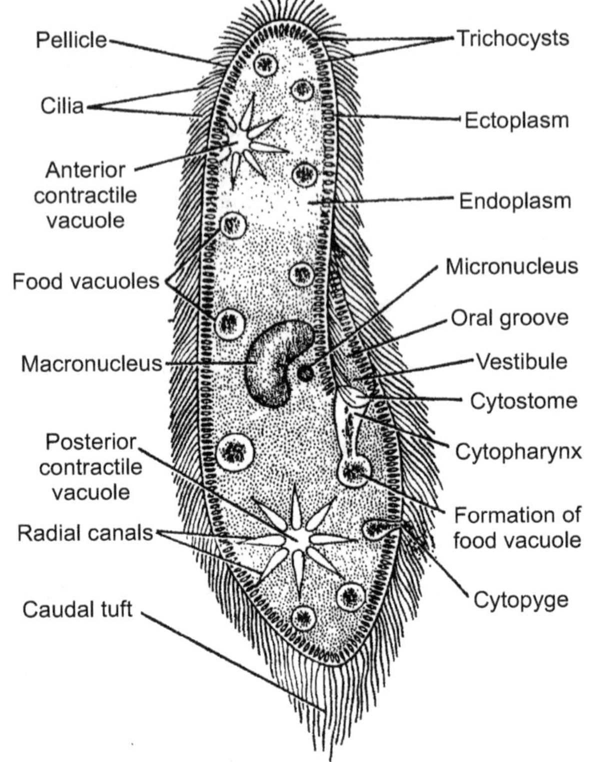

Fig. 1.2 : *Paramoecium caudatum*

6. Body is covered by cilia which are locomotory and tactile as well as derive food material in mouth.

7. Endoplasm contains reserve food vacuoles, mitochondria, golgi body, ribosomes, various crystals, nuclei, contractile vacuoles etc.

8. *Paramoecium* has two nuclei – a small rounded, micronucleus (controls reproductive activity, bounded by nuclear membrane) and a large membraneless, macronucleus (controls metabolic activity).

9. It has two contractile vacuoles with excretory and osmoregulatory functions.

10. Reproduction is asexually by binary fission (if favourable conditions) and sexually by conjugation during unfavourable conditions.

✱✱✱

3. AMOEBA

Classification :

Kingdom	:	Protista (Protozoa)
Phylum	:	Sarcomastigophora
Class	:	Rhizopodea
Subclass	:	Lobosia
Order	:	Amoebida
Genus	:	*Amoeba*
Species	:	*proteus*

Habits and Habitat :

Amoeba proteus lives in fresh water ponds, pools, ditches and streams having green plants; *Amoeba* feeds on organic matter. *Amoeba proteus* name is derived from two Greek words; Gr. *Amoeba* = change; *proteus* = a mythological sea god, who could changes its shape.

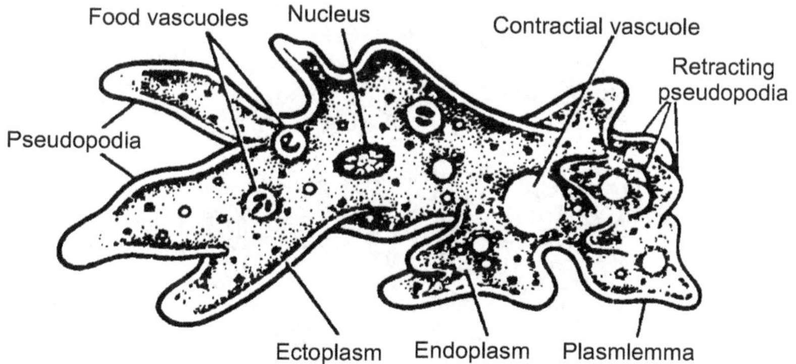

Fig. 1.3 : *Amoeba proteus*

Salient Features :

1. *Amoeba* is an unicellular and appears as colourless. The size varies from 0.2-0.3 mm in diameter.
2. It covers with plasmalemma.
3. The cytoplasm is differentiated into an outer thin cortical layer called *ectoplasm* and an inner medullary mass called *endoplasm.*
4. Nucleus, contractile vacuole, food vacuoles, water globules and other organelles such as endoplasmic reticulum, mitochondria, golgi complex, lysosomes and ribosomes are present in endoplasm.
5. Locomotion by pseudopodia.
6. Nutrition is holozoic or zootrophic and food is organic substance.
7. Respiration is aerobic and takes place through general body surface by diffusion.
8. Reproduction takes place by only asexual method i.e. binary fission.
9. *Amoeba* has a great power of regeneration.

4. *PLASMODIUM SP.*

Classification :

Kingdom	:	Protista
Phylum	:	Protozoa
Class	:	Sporozoa, Subclass - Coccidia
Order	:	Haemospordia
Genus	:	*Plasmodium*
Species	:	*Vivax*

Habit and Habitat :

It is an intracellular parasite in red blood cells of man. Its adult condition is called trophozoite. It infects birds, reptiles and mammals. There are two hosts of *Plasmodium vivax* man and female anopheles mosquito.

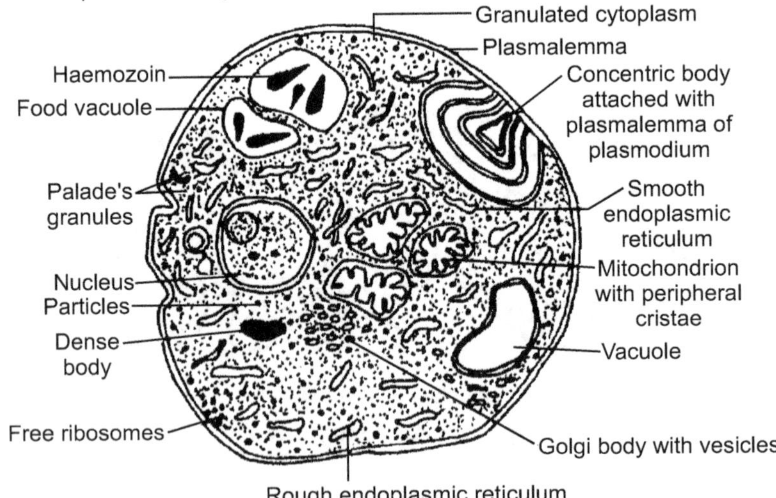

Fig. 1.4: *Plasmodium*. Ultra structureof trophozoite in R.B.C. as seen in electron microscope

Salient Features :

1. *Plasmodium vivax* causes malaria in humans.
2. It is carried to man by vector anopheles female mosquito.
3. Its life cycle is completed into two hosts man and *anopheles* female mosquito.
4. In man asexual phase of life cycle is completed called schizogony.
5. The sexual phase of life cycle is completed in female anopheles mosquito by gametogony, syngamy and sporogony.
6. The mature adult parasite of plasmodium is called trophozoite and it lives in RBCs of man.
7. The trophozoite is amoeboid in shape, uninucleated having granular and vacuolated cytoplasm.

8.   The granules are chiefly of haemozoin pigmens.

9.   Electron studies showed that the parasite has double membrane plasmalemma, mitochondira, golgi complex and nucleus.

10. It mode of nutrition is saprozoic.

11. Female anopheles injects sporozoites in the body of man. Sporozoites are sickle shaped, with oval nucleus.

12. There are 60 species *plasmodium.*

13. For the control of malaria control of mosquitoes is very important.

14. Mepacrine, Daraprim are effective drugs against *plasmodium.*

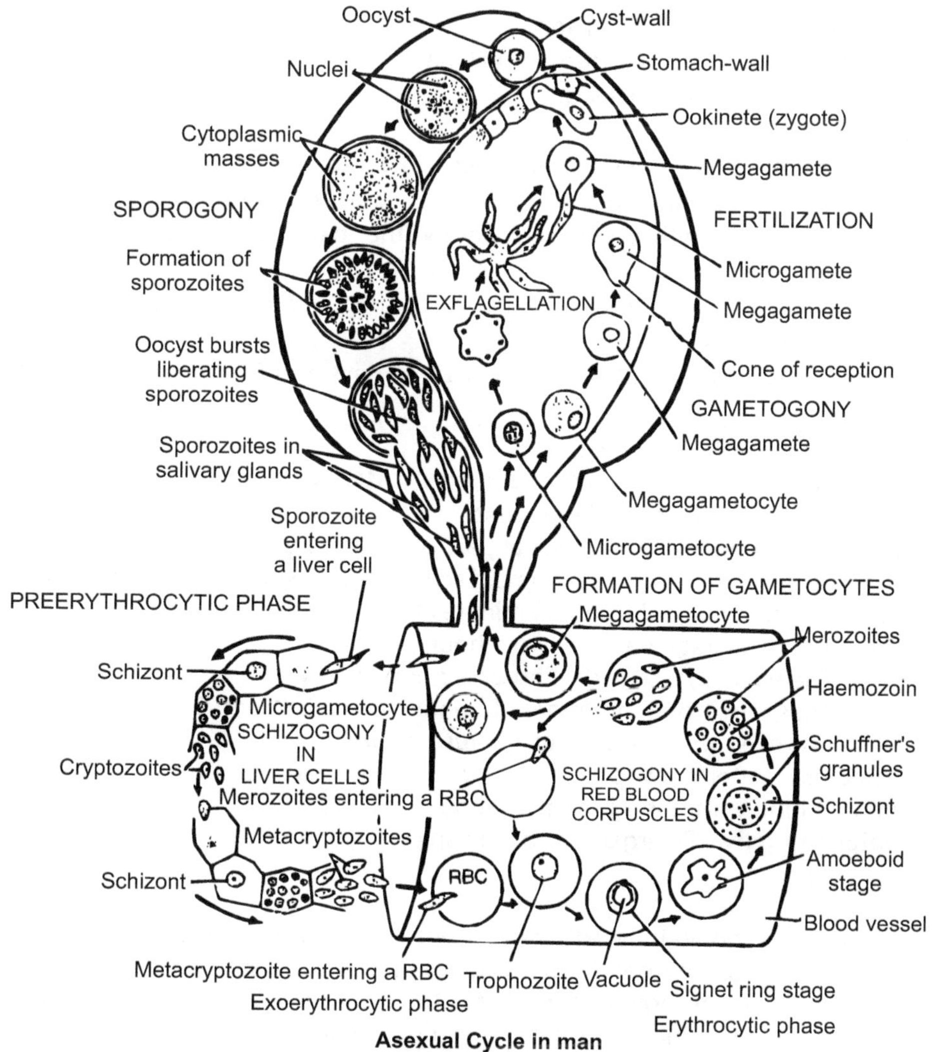

Fig. 1.5 : *Plasmodium vivax*. Life cycle

Pathogenicity :

Infection with the *Plasmodium* causes intermittent fevers which are known as malaria. The merozoites at the end of schizogony rupture the blood corpuscles and release toxic substance 'Haemozoin' in the blood. This toxin then deposited in the spleen, liver, and under the skin, so that the host gets yellow colour. The accumulated toxins cause malaria fever in which the patient suffers from chills, shivering and high temperature with convulsion followed by profuse sweating. The fever last from 6 to 10 hours, then it comes on again after every 48 hours coinciding with the liberation of new generation of merozoites.

Each of the four species causes a characteristic fever and the diseases are designated as follows :

P. vivax	:	Bengin tertian malaria.
P. malariae	:	Quartan malaria.
P. falciparum	:	Malignant tertian-malaria.
P. ovale	:	Ovale malaria.

Incubation Period :

The sporozoite after gaining entrance into the human body undergoes a developmental cycle first in liver and then in the red blood corpuscle. With the commencement of the erythrocytic schizogony the parasite multiplies in geometrical progression and on reaching a sufficient concentration in the blood, brings about the onset of fever. This period of development is called the incubation period, which varies with different species as follows :

In *P. vivax*, *P. ovale* and *P. falciparum*, it is 10-14 days and *P. malariae*, it is 18 days to 6 weeks.

Clinical Features of Malaria :

The main clinical manifestations in a typical case, are a series of febrile paroxysms followed by anaemia and splenic enlargement.

1. **Febrile paroxysms :** The malarial paroxysm starts generally in the early afternoon but actually it may start at anytime. Each paroxysm shows a succession of 3-stages:
   (a) The cold stage (lasting 20 minutes to an hour).
   (b) The host stage (lasting 1 to 4 hours).
   (c) The sweating stage (lasting 2 to 3 hours).
   Thus, the total duration of febrile cycle is from 6 to 10 hours, varying however, with the species of *Plasmodium*.
2. **Anaemia :** After a few paroxysms, anaemia develops as a result of breaking down of red blood cells during segmentation of parasites.
3. **Splenomegaly :** Enlargement of the spleen is one of the important physical signs in malaria.

Control :

1. Mosquitoes should be eradicated for better control. (Anopheles vector).
2. Man should save himself from mosquito bite.
3. Various drugs are now under use for the treatment of malaria including Quinine, Atabrin, Chloroquine, Camoquin, pamaquine, Paludrine, Daraprim, etc.

✶✶✶

Practical **2**...

Aim: To perform Museum Study of Phylum Porifera : *Sycon, Euplectella, Chalina, Spongilla.*

PHYLUM *: PORIFERA*

Sponges which include in Phylum *Porifera*, are most primitive, multicellular animals. The name Porifera (L, porous = pore; ferro = to bear) is given to this phylum by Robert E. Grant in 1836 means pore bearers.

General Characters of Phylum Porifera :

1. All sponges are aquatic, mostly marine and few are fresh water.
2. They are multicellular organisms with cellular grade of body organisation.
3. They are sessile, solitary or colonial.
4. The surface of the body is perforated by large number of pores and canals.
5. Body is *asymmetrical* or radially *symmetrical.*
6. The *osculum* is a large opening through which water is released out, no mouth and digestive system. Digestion intracellular.
7. The nervous and sensory cells are absent.
8. The body wall is *diploblastic* with outer dermal epithelium and inner gastral epithelium with a gelatinous non-cellular *mesenchyme* in between.
9. Sponges have endoskeleton of *calcarous* or *siliceous* spicules and proteinous spongin fibres.
10. All sponges possess great power of *regeneration.*
11. Reproduction takes place both by *asexual* and *sexual* method. Asexual by means of formation of buds and *gemmules.* Sexual by means of *ova* and *sperms.*
12. Fertilization is internal and cleavage is *holoblastic.*
13. Development is indirect by a free swimming ciliated larva, the *parenchymula* or amphiblastula.

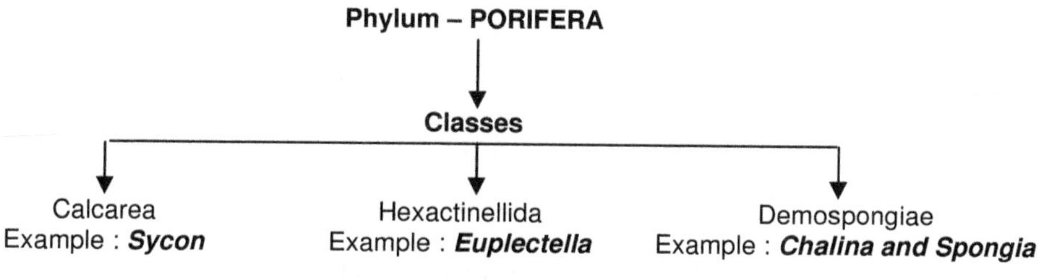

1. *SYCON*

Classification :

Phylum	:	Porifera – Diploblastic, pore bearing, cellular grade, asymmetrical or symmetrical.
Class	:	Calcarea – Endoskeleton is formed of calcareous spicules.
Order	:	Heterocoela – Canal system syconoid type.
Genus	:	*Sycon* or *Scypha.*

Habit and Habitat :

Sycon is a common solitary sponge, found attched to the rocks and to some submerged solid object in the shallow marine water.

Distribution :

Found in North Atlantic shores.

Salient Features :

1. The body is slender, vase-shaped, radially symmetrical cylinder measuring 2 to 8 cm long and 0.5 to 0.8 cm in diameter.
2. Each cylinder is diliated at the middle and opens through an *osculum* at the tip.
3. *Osculum* surrounded by *oscular fringe* composed of monaxon spicules.
4. Body wall is thick from which spicules project.
5. Surface of the body is perforated by numerous pores, called the *ostia* or incurrent pores.
6. Skeleton consist of calcareous spicules of monaxon, triaxon and tetraxon type.

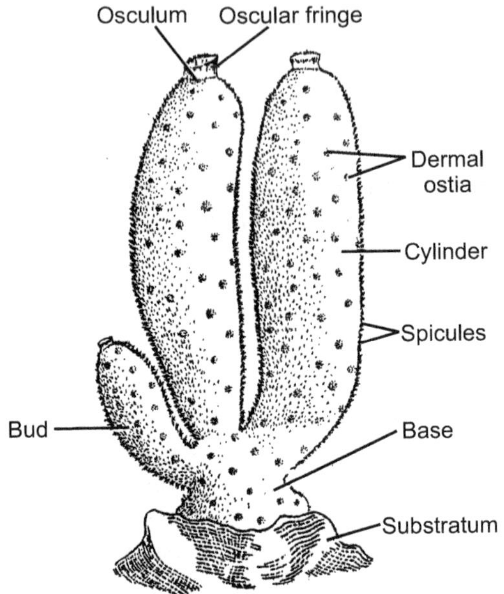

Fig. 1.6 : *Sycon or Scypha*

7. Canal system is syconoid type. Water enters the body by *ostia* and passes into the radial canals from this water reaches into the spongocoel then passes out by an osculum.
8. Reproduction both asexual by budding or regeneration and sexual by ova and sperms.

2. EUPLECTELLA

Classification :

Phylum	:	Porifera. Characters are same as *sycon*
Class	:	Hexactinelida skeleton consists of triaxon, six rayed, siliceous spicules
Order	:	Hexasterophora. Spicules are hexa asters, i.e. starlike in shape.
Genus	:	*Euplectella*
Species	:	*Aspergillum*

Habits and Habitats :

It is marine sponge found attached by its siliceous roots to the bottom of deep sea near the Philippines.

Salient Features :

1. It is commonly called as Venus' flower basket.

2. It is a glass sponge.
3. The body is cylindrical and curved with thin walls.
4. The upper end is closed by an oscular sieve formed by fused spicules and lower end anchoring siliceous root spicules.
5. Epidermis is without pinacocytes. Encircling the cylinder are projecting edges of spicules, with many openings or *parietal gaps* in the meshwork of spicules. They are connected to the spongocoel but are not the part of canal system.
6. In the wall of the sponge flagellated chambers are situated radially.
7. Skeleton is made of four and six rayed siliceous spicules bound together by siliceous cement so that they form the network of great complexity.
8. Canal system is simple sycon type.
9. In the spongocoel pair of crustacean live in commensal relationship feeding on plankton brought in with the water current.

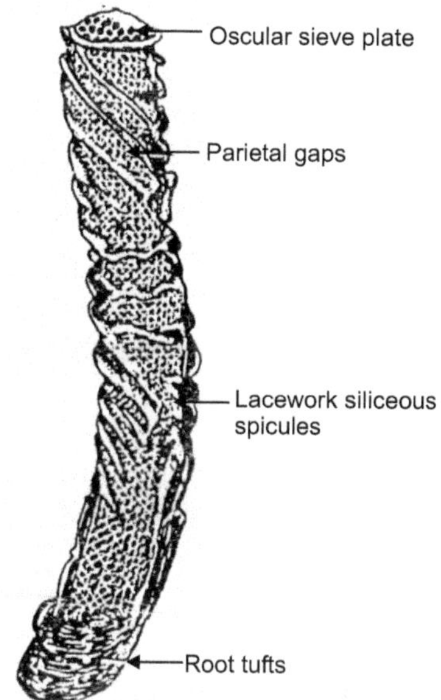

Fig. 1.7: *Euplectella aspergillum*

3. CHALINA

Classification :

Phylum	:	Porifera. Characters are same as *sycon*.
Class	:	Demospongiae. Skeleton may be of sponging fibres or sponging fibres with spicules or may be no skeleton.
Order	:	*Haplosclerida*. Spicules are of only one type called monaxon megascleres and sponging fibres are generally present.
Genus	:	*Chalina*

Habits and Habitat :

It is found in deep sea waters from Rhode island to Labrador.

Salient Features :

1. It is popularly known as 'dead man's finger', because it is shaped like a hand with many fingers.
2. Each branch or finger like structure is perforated by numerous oscula.
3. Skeleton consists of sponging fibres with siliceous monoaxon type of spicules.
4. It shows reproduction by both asexual and sexual method. Asexually regeneration and budding, sexual by producing free swimming larva.
5. The sponge shows orange, red or yellow colour.
6. Not found in shallow waters.

Fig. 1.8: *Chalina*

4. *SPONGILLA*

Classification :

Phylum	:	Porifera. Characters are same as sycon.
Class	:	Demospongiae. Skeleton may be of spongin fibres with siliceous spicules.
Order	:	Hapiosclerida. Spicules are of only one type called monaxon megascleres and spongin fibres are also present.
Genus	:	*Spongilla*

Habit and Habitat :

It is best known freshwater, sponge found in ponds, lakes and slow streams. It grows on submerged sticks and plants.

Salient Features :

1. It is freshwater sponge.
2. It forms colony and it is profusely branched.
3. Sponge shows various shades of green colour due to presence of *Zoochlorellae*, a green algae in the tissues.
4. The body wall consists of dermal membrane with dermal pores and several oscula.
5. It has rhagon type of water canal system.
6. Skeleton consists of siliceous spicules in the form of network of smooth or spiny large and small *oxeas* embedded in the spongin.
7. Reproduction both asexual and sexual.
8. Asexual reproduction by gemmules.
9. Sexual reproduction by way of unusual free swimming larva which is characteristic of *spongilla*.

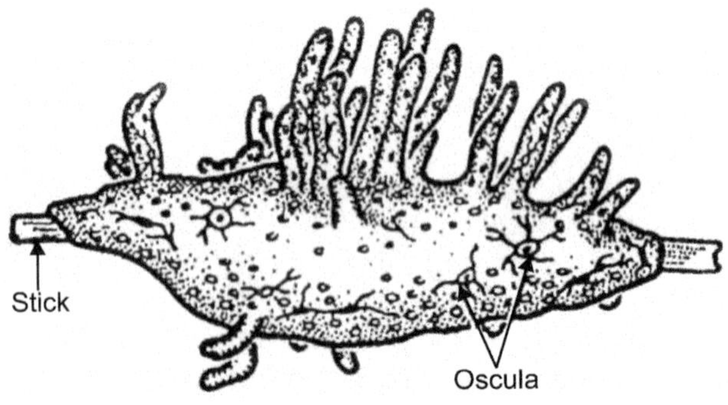

Fig. 1.9: *Spongilla*

***

Practical 3...

Aim : To do a Museum Study of Phylum Cnidaria. *Hydra, Physalia, Aurelia, Metridium.*

PHYLUM – COELENTERATA

The term coelenterata was first used by Leukart in 1847 for those animals in which the enteric cavity form the body cavity. The name derived from two Greek words, Koilos = hollow, Enteron = intestine.

General Characters :

1. The animals are radially symmetrical and diploblastic with only ectoderm and endoderm. The jelly like mesoglea is present in between these two layers which is structureless.
2. The ectoderm and endoderm form poorly organised body tissues.
3. The radially symmetrical body has a single *coelenteron* or *gastrovascular* cavity which has only one aperture called mouth.
4. The gastrovascular cavity is lined by endoderm; ceolom is absent.
5. They bear tentacles which possess characteristic nematocysts.
6. Nervous system is primitive consisting of *neurons* which exhibit network.
7. Most of the coelenterates are marine except few are fresh water.
8. They have two different structural types called *polyps* and *medusae*.
9. Asexual reproduction by budding and in sexual reproduction there is an oval ciliated larva called *planula* during development.

The phylum coelenterata is divided into three classes :

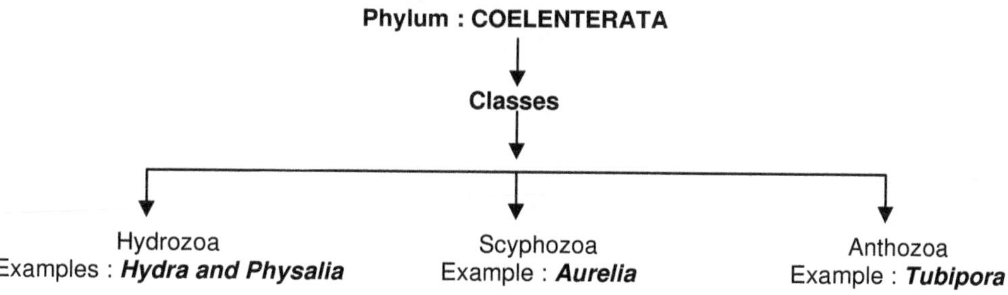

1. *HYDRA*

Systematic Position :

Phylum	–	Coelenterata	– Tissue grade; diploblastic and acoelomate
Class	–	Hydrozoa	– Hydroids bearing medusa with velum
Order	–	Hydroida	– Polypoid generation well developed
Sub-order	–	Anthomedusae	– Hydrotheca and gonotheca are absent
Genus	–	*Hydra*	– Hydrotheca and gonotheca are absent

Habit and Habitat :

It is a fresh water solitary animal, found attached to stones, rocks and weeds in ponds, lakes etc.

Distribution :

It has cosmopolitan in distribution, but most common in India, Canada and U.S.A.

Salient Features :

1. Body of *Hydra* is cylindrically elongated, elastic and tube-like measuring 1 to 3 cm in length.
2. Proximal end is closed and known as *basal disc* or foot meant for attachment to the substratum.
3. Free distal end bears a *mouth*. It is situated on a conical elevation called *hypostome.*
4. The hypostome is encircled by 6-10 tentacles. The tentacles are hollow, filiform structures provided with *nematocysts.*

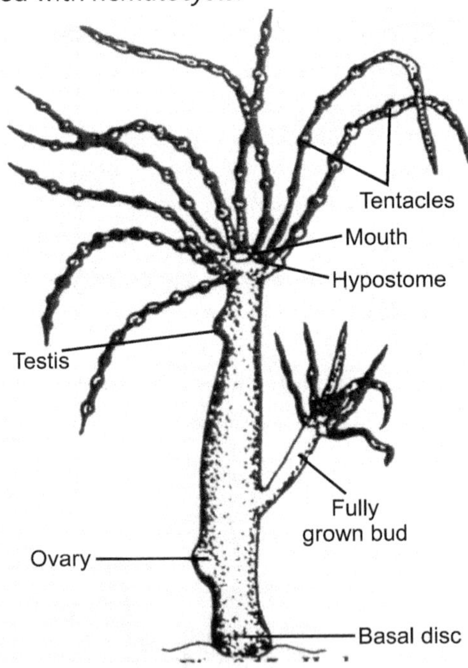

Fig. 1.10 : *Hydra*

5. Body is diploblastic, consisting of outer ectoderm and inner endoderm separated by mesogloea.
6. Body wall encloses a gastro-vascular cavity or coelenteron extending into tentacles.
7. Reproduction both asexual and sexual. Asexual reproduction by budding. Lateral buds develop on the sides of the body which later on detach and develop into a new *Hydra.*
8. Sexual reproduction by fusion of gametes. Gonads appear as buds on the sides of body. Testis lies near the oral end and ovaries near the base.

9. Development direct without larval stage.

10. *Hydra viridis* is green *Hydra*. It contains symbiotic green algae, *Zoochl*

11. *Hydra* has great experimental value as regeneration and grafting ex performed on it.

2. *PHYSALIA*

Classification :

Phylum : Coelenterata – Multicellular, diploblastic, acoelomate, nematocysts.

Class : Hydrozoa – Fresh and marine water forms, polymorphic with true velum.

Order : Siphonophora – Free swimming colonies, perisarc is absent, show extreme polymorphism in polyps and medusae.

Genus : Physalia

Habit and Habitat :

It is found in the gulf stream from Florida to Vineyard and occassionally to the Bay of Fundy, Floating or swimming.

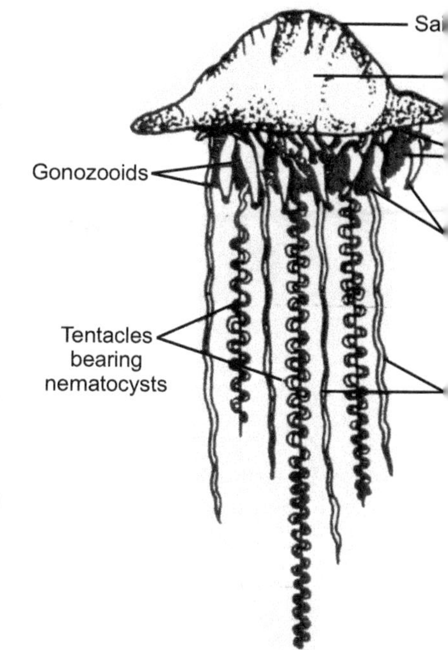

Fig. 1.11 : *Physalia*

Salient Features :

1. *Physalia* is a colonial hydroid commonly known as *Portuguese man of*

2. It is a marine, pelagic, polymorphic hydroid.

3. Body is elongated, bladder-like, gas-filled pneumatophore, narrow a and with a dorsal crest or sail.

4. The swimming bells or nectocalyces absent.

5. From the underside of the penumatophore hang dactylozooids wi gastrozooids without tentacles and branching blastostyles.

6. Tentacles are refractile.

7. Physalia is measuring about 20 to 30 cm in length, and 10 cm in width.

3. AURELIA (JELLY FISH)

Classification :

Phylum	:	Coelenterata – Multicellular, diploblastic, acoelomate, nematocyte.
Class	:	Scyphozoa – Adult medusoid; gastric filaments endodermal; true vellum absent, oral arms present, gonads endodermal, tentaculocysts present.
Order	:	Semacostomeae – Flattened, disc-shaped medusa, square mouth guarded by four long oral arms; tentaculocysts usually eight, gonads subgstral.
Genus	:	*Aurelia*.

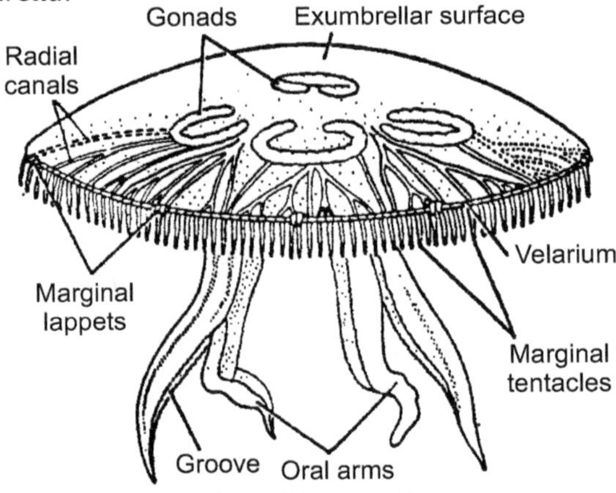

Fig. 1.12 : Aurelia

Habit and Habitat :

    Aurelia is found in coastal waters of all oceans of the world. They are solitary, marine jelly-fish.

Salient Features :

1. Aurelia commonly called *jelly-fish*.
2. Umbrella-like *medusa* with a convex exumbrellar and a concave subumbrellar surface.
3. Subumbrella contains marginal tentacles. 8-marginal lappets and 8-tentaculocysts.
4. A pendent *manubrium* hangs from the centre of the subumbrellar surface.
5. Oral arms-4, long and perradial in position.
6. Mouth leads to a short *gullet*, which opens in the stomach.
7. Stomach sends - 4-interradial gastric pouches.
8. Radial canals subdivided into branches ending in the circular canal.
9. Gonads horse-shoe shaped.
10. A ciliated exhalant groove present on the lower side of the oral arm.
11. Unisexual.

✱✱✱

4. SEA ANEMONE

Classification :

Phylum	:	Cnidaria – Soft bodied, multicellular, diploblastic, acoelomate, nematocysts.
Class	:	Anthozoa – Only polyps, solitary or colonial, symmetry polymerous, enteron divided into compartments by mesenteries, marginal tentacles around the mouth.
Subclass	:	Hexacorallia – Tentacles numerous, simple and branched.
Order	:	Actiniaria – Simple or colonial, skeleton absent, tentacles and mesenteries are arranged in multiples of six.
Genus	:	*Metridium*
Species	:	*senil*

Habit and Habitat :

The Sea-anemone is a marine polyp. It occurs commonly on rocks, weeds, shells etc. It is found from New Jersey, Coast of Europe. It feeds on plankton.

Salient Features :

1. *Metridium senil* is commonly called sea anemone.
2. Body is short, cylindrical, divided into pedal disc, column and oral disc.
3. It measures 5-7 cm in length, body is brightly coloured.
4. *Pedal disc* for attachment to the substratum. *Oral disc* – flattened and crowned with marginal tentacles around the mouth.

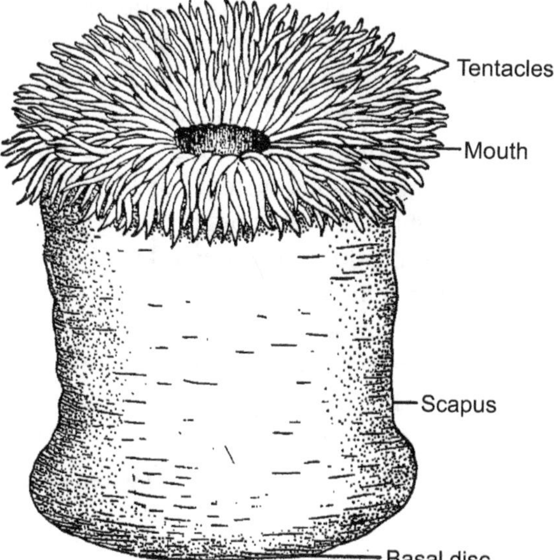

Fig. 1.13 : Metridium senil (Sea anemone)

5. Mouth leads into a short gullet with siphonoglyphs.
6. Gullet opens in the gastrovascular cavity which is divided into compartments by six pairs of mesenteries.
7. Sea anemone is voracious and feeds on molluscs, crustaceans, worms etc.
8. Respiration is aerobic.
9. Reproduction takes place asexually as well as sexually.
10. Metridium has a power of regeneration.

***

Practical **4**...

Aim : **To do a Museum Study of Phylum Platyhelminthes** *Planeria, Fasciola hepatica,*
Taenia solium

PHYLUM : PLATYHELMINTHES

The name platyhelminthes (Greek; Platy = broad or flat; helminthes = worm) means
"flat worms" which refer to their characteristics contour of flattened body.

General characters :

1. Body is soft, bilaterally symmetrical and dorsoventrally flattened but not
   metamerically segmented.
2. They may be free living or parasitic.
3. Anterior end of the body forms definite head.
4. They are triploblastic, acoelomate animals.
5. Hooks, spines, thorns, spicules and teeth are present, made of hard cuticle.
6. Digestive system and anus is absent in some forms, while in some mouth, pharynx
   and intestine is present.
7. Excretory system consists of protonephridia.
8. The nervous system consists of a network, but it has a ganglia at the anterior end
   which serves as brain.
9. Reproductive organs are highly developed and most of the flatworms are
   hermaphrodite.
10. Fertlization is internal.
11. Life cycle involves one or more hosts.

About 10 to 15 thousand species of flatworms have been described and they are
grouped in three classes.

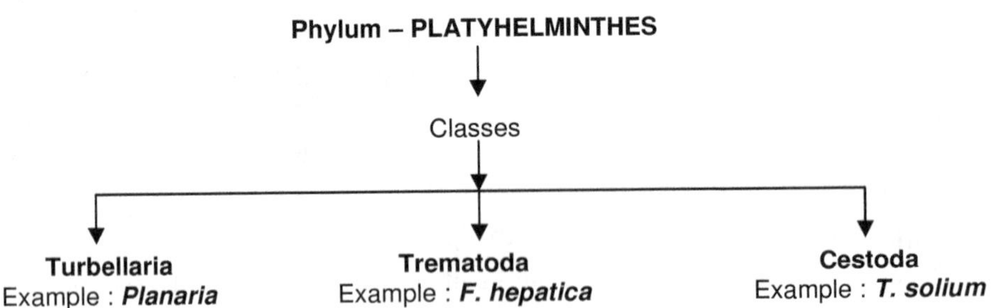

1. *PLANARIA*

Classification :

Phylum	:	*Platyhelminthes* – Dorso-ventrally flattened, acoelomate, hermaphrodite.
Class	:	*Turbellaria* – Body unsegmented, free living, epidermis contains rhabdites and covered with cilia.

| Order | : | *Tricladida* – Pharynx plicate usually directed backwards, mouth ventral, genital aperture single. |
| Genus | : | *Planaria* (*Dugesia*). |

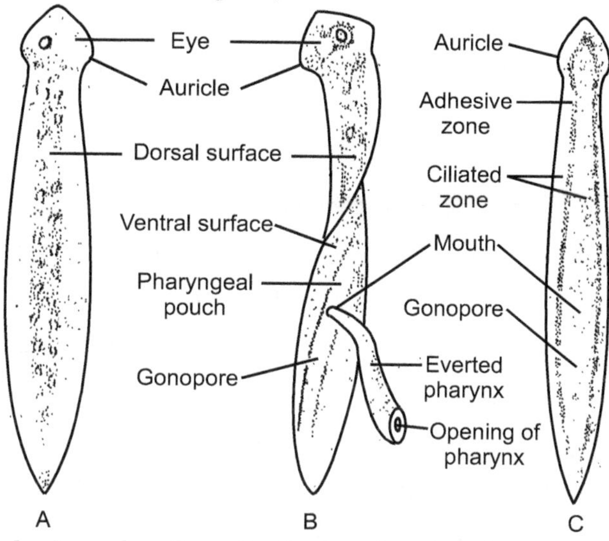

Fig. 1.14 : *Planaria.* **A – Dorsal surface; B – Body twisted to show a part of ventral surface; C – Ventral surface**

Habit and Habitat :

Dugesia is a freshwater triclad; found in streams, ponds, lakes springs under the stones and leaves throughout the temperate zones; particularly in India, Myanmar, U.K., U.S.A. and U.S.S.R.

Salient Features :

1. Body elongated, dorso-ventrally flattened and is divisible into head and body.
2. *Dugesia* are 2-15 mm in length and brown to black in colour.
3. Body tapers posteriorly to a pointed end.
4. Head is triangular with two ear-like auricles and two eyes.
5. Ventral mouth surrounded by proboscis pore.
6. Proboscis highly muscular and enclosed in a sheath.
7. Pharynx plicate usually directed backwards.
8. Intestine divided into three branches, one median and two lateral.
9. Genital pore is situated posterior to the mouth.
10. Reproduction sexual, asexual and by regeneration.

2. FASCIOLA HEPATICA (LIVER FLUKE)

Systematic Position :

Phylum	-	Platyhelminthes	-	Acoelomate; organ grade flatworms
Class	-	Trematoda	-	Flukes, ectoparasites or endoparasites
Order	-	Digenea	-	Endoparasites; mostly with two suckers
Genus	-	*Fasciola*		
Species	-	*hepatica*		

Habit and Habitat :

It is a digenetic parasite found in the bile duct and biliary passages in the liver of sheep.

Distribution :

It is cosmopolitan in distribution throughout sheep rearing areas. It is endemic in India, China, USA and Argentina etc.

Salient Features :

1. *Fasciola* is commonly known as *Liver fluke*.

2. Body is leaf-like, dark brown and dorsoventrally flatenned. The anterior end is produced into a conical projection called *cephalic cone*.

3. Mouth is present at the anterior end, surrounded by oral sucker.

4. A large highly muscular ventral sucker or *acetabulum* is present in a little behind the oral sucker.

5. Between the oral and ventral suckers, there is a *gonopore* on ventral side.

6. An excretory pore is present at the posterior extremity of the body.

7. Alimentary canal is incomplete. It consists of mouth followed by muscular pharynx, oseosphagus and a bifid intestine with diverticula.

8. Hermaphrodite. The male reproductive organs are a pair of testis, vasa deferentia, seminal vesicle, ejaculatory duct and cirrus or penis. The female reproductive organs are single ovary, oviduct, uterus and vitelline glands.

9. Life cycle is completed in two hosts, sheep (primary host) and a mollusc *Lymnaea* (intermediate host).

10. The life history stages include zygote → miracidium → sporocyst → redia → cercaria → metacercaria → adult parasite.

11. It causes a disease called *liver rot* in sheeps.

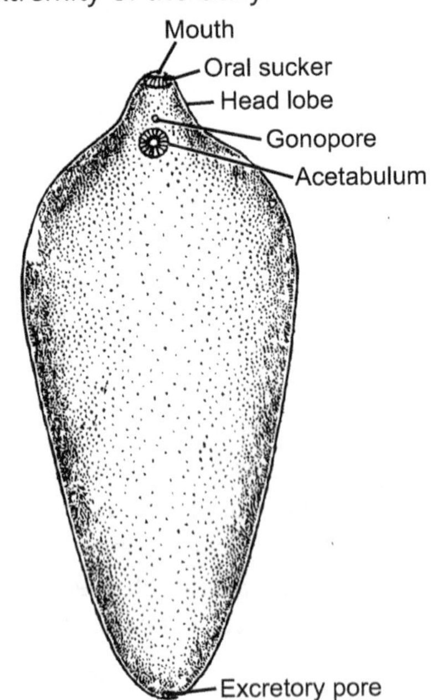

Fig. 1.15 : *Fasciola* hepatica

3. *TAENIA SOLIUM*

Classification :

Phylum	:	*Platyhelminthes* – Characters are same as *Dugesia*.
Class	:	*Cestoda* – Endoparasitic in the intestine of vertebrate, flattened body divisible into segments (proglottids) and small head with hooks and suckers, larva free swimming and hexacanth.
Order	:	Taenioidea or cyclophyllida. *Scolex* bears four suckers and a terminal rostellum with a crown of hooks. Each proglottid contains a complete set of reproductive organs, yolk gland is single and compact.
Genus	:	Taenia
Species	:	*solium*

Habit and Habitat :

Taenia solium is commonly found in the intestine of man in places where pork is eaten as food. It is cosmopolitan in distribution.

Salient Features :

1. It is commonly known as tapeworm.

2. Body is divided into *scolex, neck* and *strobilla*, yellowish white in colour.

3. Scolex bears four cup-shaped *suckers* and a terminal *rostellum* crowned with double rows of hooks.

4. Hermaphroditic, sex organs are fully developed in posterior segments.

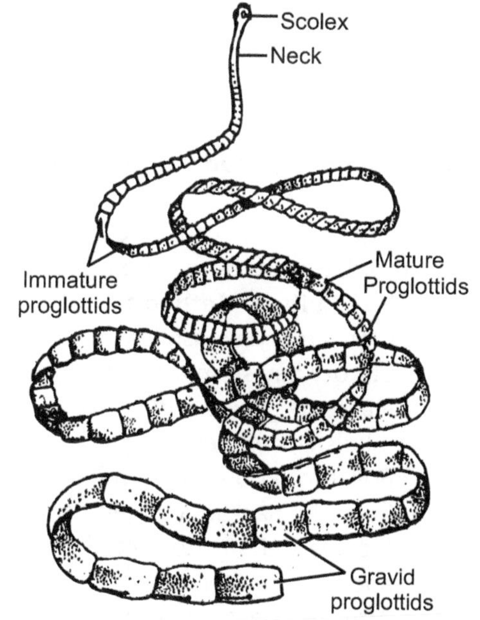

Fig. 1.16 : *Taenia solium*

5. Ripe uterus with a central stem and lateral branches, usually 7 to 12 on each side.

6. Life cycle consists of intermediate host pig and definite host man.

7. *T. Solium* causes disease *Taeniasis.*

✱✱✱

Practical **5**...

Aim : To Study *Paramoecium* : Culture External Morphology, Conjugation and Binary Fission.

(a) Aim : To setup a *Paramoecium* culture (E).

Requirement :

Glass jar, conical flask, pond water, moisten wheat grains etc.

Collection :

Paramoecium is a very active, free swimming protozoa. They are common in fresh water with decaying vegetable matters and abundance of bacteria. *Paramoecia* appears to be small, white, slightly elongated, fast moving object. They multiply rapidly in water containing decaying organic matters and high bacteria population. Collect the pond water along with the decomposing weeds.

Culture of *Paramoecia* :

1. **Simple Method :** Collect submerged fresh and decomposing weeds with a small amount of water from a pond in which *Paramoecia* are present. Mix all those in distilled water in a glass jar, cover it and leave for few days (2 to 4 days). The weeds decompose within a few days and the culture is rich in *Paramoecium*.

2. **Hay Culture Method :**
- Moisten and smash 20 wheat grains.
- Cut dry hay stems (weed) into small pieces.
- Place the grains and 20 to 25 pieces of hay stem in a conical flask containing 500 ml distilled water.
- Boil it for about 10-15 minutes.
- Keep the flask in a dry and cool place for about four days.
- Inoculate the culture medium with *Paramoecia* already available in simple method and wait for few days.
- After few days numerous *Paramoecia* will grow in the culture.
- Use the same culture for further studies i.e. External morphology, different organelles, behaviour and response to various stimuli.

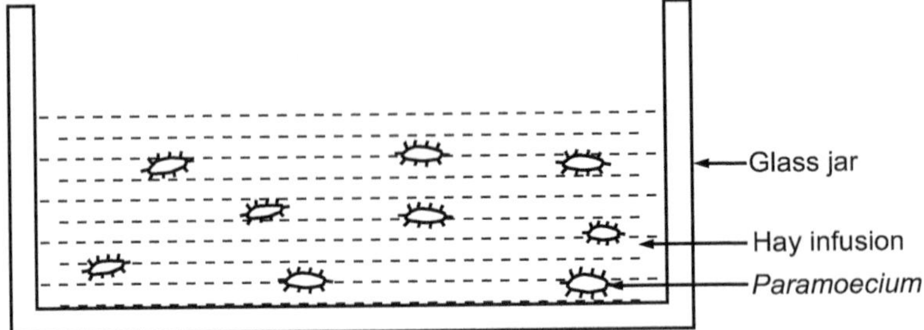

Fig. 1.17 : *Paramoecium* culture

(b) Aim : To study external characters of *Paramoecium.*

Systematic Position :

Kingdom	–	Protista
Subkingdom	–	Protozoa
Phylum	–	Ciliophora
Class	–	Ciliata
Subclass	–	Holotrichia
Order	–	Hymonostomatidia
Family	–	Paramecideae
Genus	–	*Paramoecium*
Species	–	*caudatum*

External Characteristics of *Paramoecium* :

1. *Paramoecium* are found in ponds, lakes, ditches, stagnant water, river, etc.
2. It is a microscopic animal measuring about 200 to 300 microns in length.
3. It feeds on tiny protozoans, bacteria, bites of animal.
4. *Paramoecium* looks like the sole of slipper and that is why the animal is also called "slipper animalcule".
5. It is streamlined, asymmetrical with well marked oral and aboral surfaces on the body.
6. Body cytoplasm divides into ectoplasm and endoplasm. Ectoplasm secretes pellicle which gives definite shape to the body along with protection being gelatinous and elastic in nature.

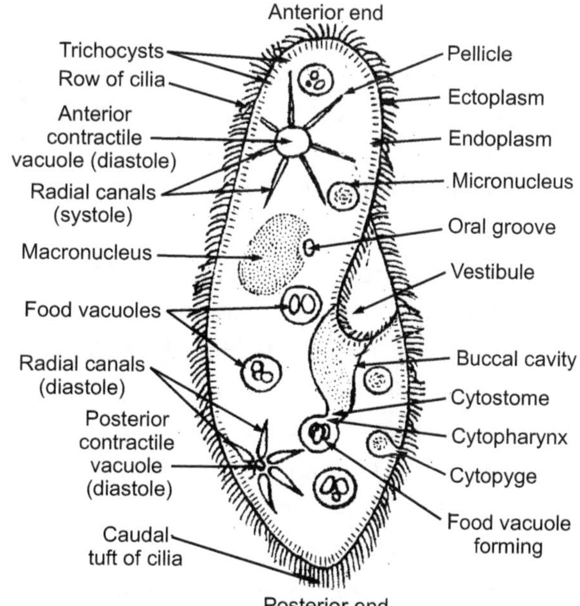

Fig. 1.18 : W. M. *Paramoecium species*

7. The body is covered by cilia which are locomotory and tactile as well as derive food material in mouth.
8. Endoplasm contains reserve food vacuoles, mitochondria, golgi body, ribosomes, various crystals, nuclei, contractile vacuole etc. It has two nuceli - a small rounded - micronucleus (controls reproductive activity, bounded by nuclear membrane) and a large membraneless macronucleus (controls metabolic activity).
9. It has two contractile vacuoles with excretory and osmoregulatory functions.
10. Reproduction is asexually by binary fission (if favourable conditions) and sexually by conjugation during unfavourable conditions.

(c) Aim : To study the Reproduction in *Paramoecium* (D).

Reproduction occurs in two mode i.e. *Binary fission* and conjugation.

(A) Binary Fission (Asexual reproduction) :

It is the division of parent body into two nearly equal daughter individuals.

1. At first *Paramoecium* stops feeding and oral groove starts disappearing.

2. The macronucleus divides amitotically, elongates and constricts in the middle.

3. Micronucleus divides mitotically but the nuclear membrane remains intact resulting two daughter micronuclei move towards opposite ends of the cell.

4. The transverse constriction appears round middle of the oblonge body and continues to grow deeper ultimately dividing the cytoplasm.

5. The body divides into two daughter *Paramoecia*, the anterior is called *proter* while the posterior as *ophesta*.

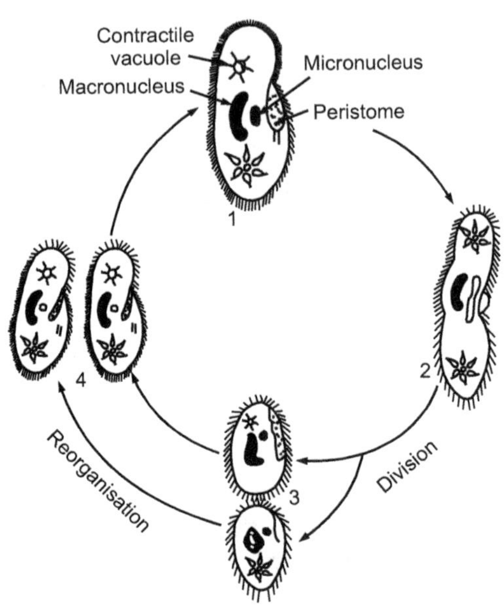

Fig. 1.19 : *Paramoecium* species - Binary fission

6. Each daughter *Paramoecia* receive one contractile vacuole from the parent and form the second - *de-Novo*.

7. Oral groove develops in each while buccal structure becomes reconstituted.

8. The two daughter *Paramoecia* are of equal size and each with a complete set of cell organelles.

9. This whole process is completed in 30-120 minutes depending on the temperature and food availability and also the salinity and pH.

(B) Conjugation (Sexual Reproduction) :

It is the temporary union of two individuals of same species for exchanging a part of their micro-nucler material necessary for continued vitality and vigour of the species.

1. Conjugation constitutes the sexual part of reproduction.

2. The two individuals come in contact for mating and unite by their own grooves.

3. The pellicle between the two forms is disintegrated, but they swim together and are called conjugants or gametocytes.

4. In each conjugant macronucleas disappears and micronucleus divides twice forming four haploid daughter micronucleii.

5. Three out of the four daughter micronucleii disintegrate in each conjugant, while remaining divides into two unequal daughter pronucleus. The smaller one is called *active potential male migratory pronucleus*, while the larger is called *stationary female pronucleus*.

6. The migratory male pronucleus of one conjugant moves through the protoplasmic bridge into other conjugant and fuses with stationary female pronucleus to form zygote and vice-versa.

7. *Zygote* nucleus is diploid and is called as synkaryon or amphinucleus. This kind of complete union of two nuclei of different individuals is known as *amphimixis*.

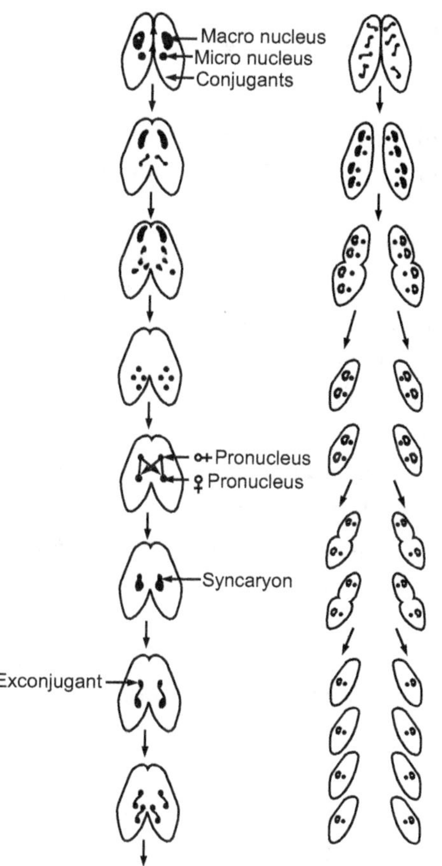

Fig. 1.20 : *Paramoecium* conjungtion

8. After this a series of nuclear division continues and after a few changes four adult *Paramoecia* are formed.

9. The process of conjugation is conditioned as it occurs in deficiency of nutrition, at a certain temperature and forms smaller individuals.

10. Conjugation occurs in different strains. It has important significance, as it imparts rejuvenation and nuclear reorganization and also facilitates in transference of hereditory materials. It also helps in the formation of better resistance *Paramoecia*.

✳✳✳

Practical **6**...

Aim : (A) To Study of permanent slides : Spicules and gemmules of sponges, T. S. of sycon, T.S. of hydra, *Taenia solium* scolex and gravid proglottid.

1. SPICULES

Skeleton of sponges is composed of spicules or spongin fibers or of both. The spicules are needle-like shiny structures. They have a central axis of organic substances around which inorganic substances are accumulated. An organic sheath is present outside this layer of inorganic substances. The organic substances of the spicules is either calcium carbonate or silica.

According to size the spicules are categorized into *megascleres* and *microscleres.*

[A] Megascleres :

These are larger spicules and named by adding *axon* in the number of their axes and *actine* or *actinal* in the number of their rays. They are of the following types :

1. **Monaxon spicules :** These are simple, straight or curved, needle-like, rod-shaped spicules, having only one axis. They are further having following kinds :

    (a) Monactinal monaxon : When the growth of the spicules take place at one of the axis, it is called *style.* Its round end is called *strongylote* and pointed end is known as *axeote.*

        (i) Tylostyle : Its round end is knob-like.

        (ii) Acanthostyle : Many minute spines are found on its surface.

    (b) Diactinal monaxon : When the growth occurs at both the ends of a monaxon spicule it is known as *diactinal monaxon* or *diactine rhabd.*

    (i) Oxeas : Their both the ends are pointed.

    (ii) Tornotes : Their ends are like spear heads.

    (iii)Strongyles : Their both the ends are round.

    (iv) Tylotes : Their both the ends are like pin-heads.

2. **Triaxon spicules :** These spicules occur only in the Hexactinellida hence also called hexactinal spicules. Each spicule consists of 3-axes crossing at right angles. Thus, six rays extend from a central point at right angles to each other.

3. **Tetraxon spicules :** These are also called *tetractines* and *quadriradiates.* Each spicule has 4-rays; and not in the plane. They are of the following types :

    (i) Calthrops : When 4-rays are more or less equal.

    (ii) Triaenes : When 1-ray is elongated (rhabdome), bearing 3-smaller rays (cladome). If one smaller ray is lost it become *diaene.*

    (iii)Amphidisc : When disc occur at both ends of a rhabdome.

4. **Polyaxon spicules :** These spicules have several equal rays radiating from a central point.

5. **Spheres :** These are round spicules in which growth takes place a centre. Also called *Desma.*

[B] Microscleres :

These are minute flesh spicules which are scattered in the mesenchyme. They are like megascleres in structure but small in size. They are of the following kinds :

1. Monaxon microscleres.
2. Polyaxon microscleres.

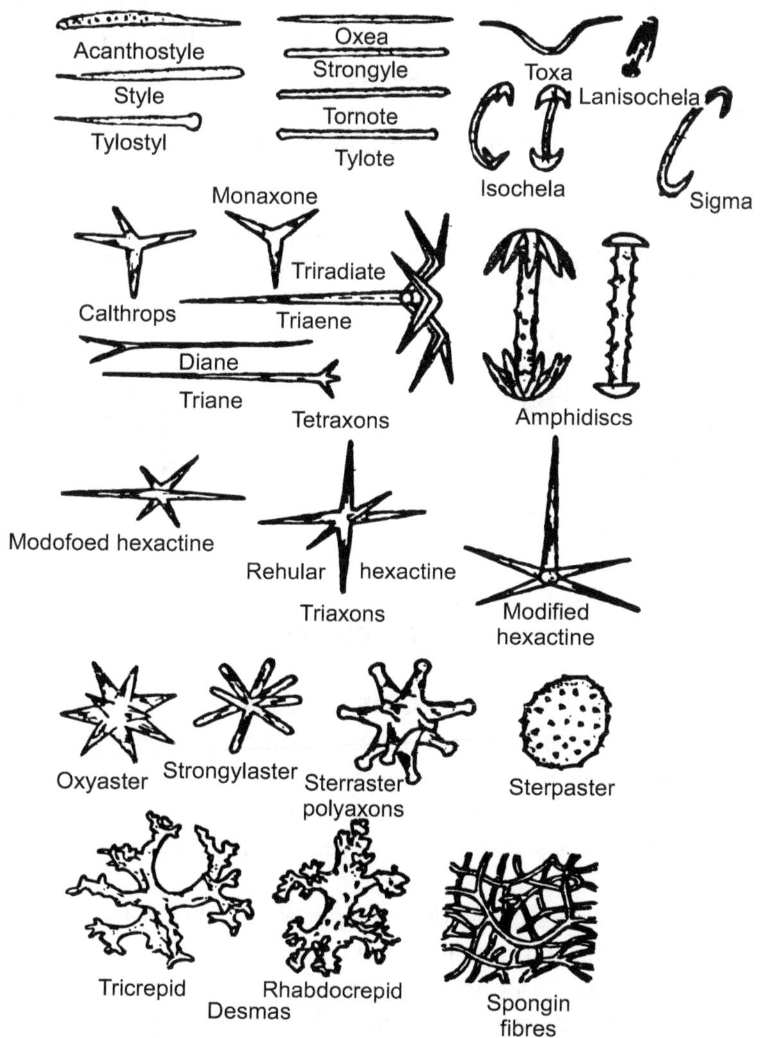

Fig. 1.21 : Spicules and spongin fibres in Sponges

2. GEMMULES

During unfavourable conditions some sponges show asexual reproduction by the development of capsulated bodies called *gemmules*. Gemmule formation takes place by the gathering of amoebocytes probably the archeocytes into a ball. This ball is surrounded by a chitinous shell supported or strengthened by spicules. *Archeocytes* are filled with food reserves constituting of glycoprotein or lipoprotein.

Gemmules possess a peripheral layer having air cells, which are helpful for floating them on the surface of the water. The spicules in the gemmules are amphidisc type. The gemmules are produced by fresh water sponges, when they are unable to live due to the excessive cold or drought. The freshwater sponges form a large number of gemmules in autumn and then disintegrate. Gemmules can withstand freezing and drying. In India, gemmules are generally found on the approach of summer. On the return of favourable conditions in the spring gemmules hatch. Archeocytes modified into scleroblasts producing spicules and after a week by differentiation and arrangement, a new complete sponge is formed.

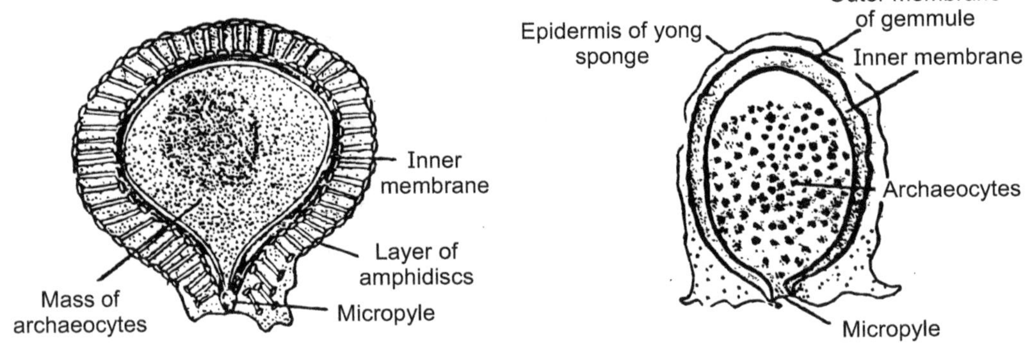

(a) Gemmule of a sponge in section   **(b) Germinating gemmule of a sponge**

Fig. 1.22 : Gemmules

3. T.S. OF *SYCON*

1. In sponges there is cellular grade of organisation means the body is composed of different types of cells lacking proper organisation.
2. The cells in sponges do not organise into tissues.
3. The T.S. of *sycon* shows distinct two layers. Outer ectoderm and inner coanoderm endoderm.
4. The ectodermal cells are called dermal epithelium. They form a covering of the outer body surface and the lining of incurrent canals.
5. The outer body cover is formed by Pinacocytes which are large, flat and closely cemented cells with thickened central bulge containing the nucleus. In *sycon*, they also line the incurrent canals and spaces found inside the body. When line the paragastric cavity they are called gastral epithelium. These cells are highly contractile and thus can reduce the surface area.

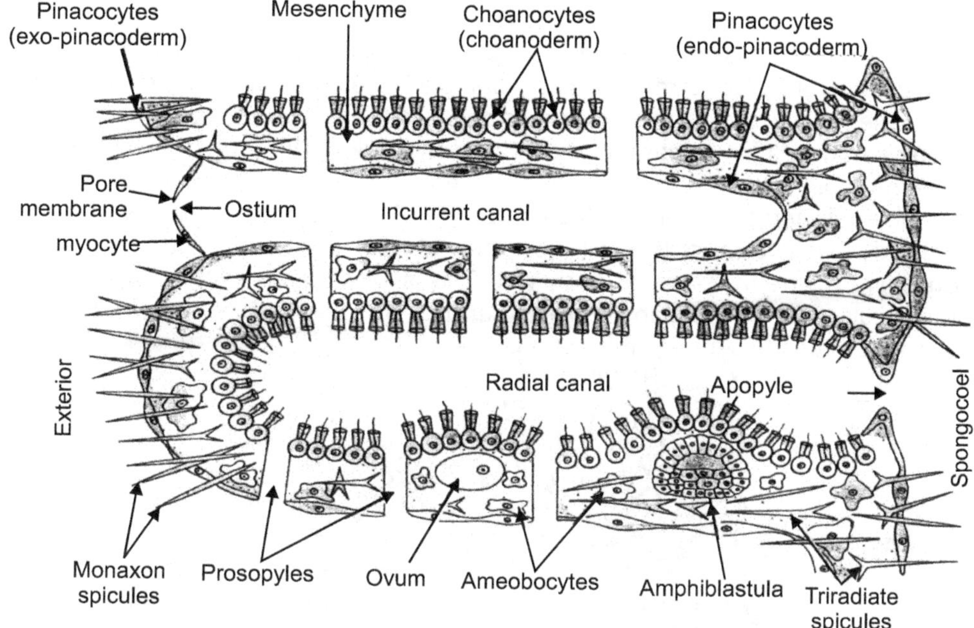

Fig. 1.23: *Scypha*. A diagrammatic sectional view of the
body wall showing one incurrent and one radial canal

6. *Pinacocytes* surrounding the osculum, outer or dermal ostia and inner ostia or apopyles are elongated and contractile and act as muscle cells called *myocytes*. They form spincters around them to regulate these openings.

7. Choanoderm or endoderm forms the gastric epithelium and consists of flagellated collar cells or *choanocytes*. They are oval or rounded cells arranged upon the mesenchyme. Each cell contain a single nucleus, one or two contractile vacuoles, food vacuoles, reserve food, blepharoplast, rhizoplast and single basal granule or kinetosome from which whip like flagellum arises. The flagellum is surrounded by at its base by a thin cytoplasmic collar. The coanocytes line the radial canal or flagellated chambers. The movement of their flagella causes a current of water.

8. Monoxon, triaxon spicules are present.

9. The central cavity is called spongocoel lined by pinacocytes.

10. In the T.S. of *scypha sycon* type of canal system shows pores and canals.

11. Dermal pores or ostia are present on the ectoderm with contractile cells called myocytes.

12. The incurrent canals are communicated with radial canals by prosopyles.

13. The radial canals are relatively wide and are polygonal in cross section.

14. The excurrent canal opens into the spongocoel through wide gastrial ostium.

15. Spongocoel is lined by pinacocytes.

16. Mesenchyme fills the space between the two body layers is composed of as thin gelatinous matrix which contains variety cells like collencytes, chromocytes, thesocytes, scleroblast, trophocytes, phagocytes, archaeocytes, lophocytes, desmocytes, gland cells, germ cells and myocytes.

4. T.S. OF *HYDRA*

Transverse section passing through the body of *hydra* shows the outermost layer of ectoderm and inner endoderm. *Hydra* is diploblastic animal and in between ecto and endoderm. They are also called as epidermis and gastroermis. These both layers are joined together by a non-cellular gelatinous layer called mesoglea. The innermost circular cavity is called coelenteron.

Both epidermis and gastrodermis are composed of different types of cells. The epidermis is made up of small cubical cells and is covered with a delicate cuticle. It forms a thin layer, about one-third of the thickness of body wall. This layer contains several types of cells like, epitheliomuscular, interstitial, gland, ciblasts, sensory, nerve and germ cells. The epidermis is protective, muscular and sensory in function.

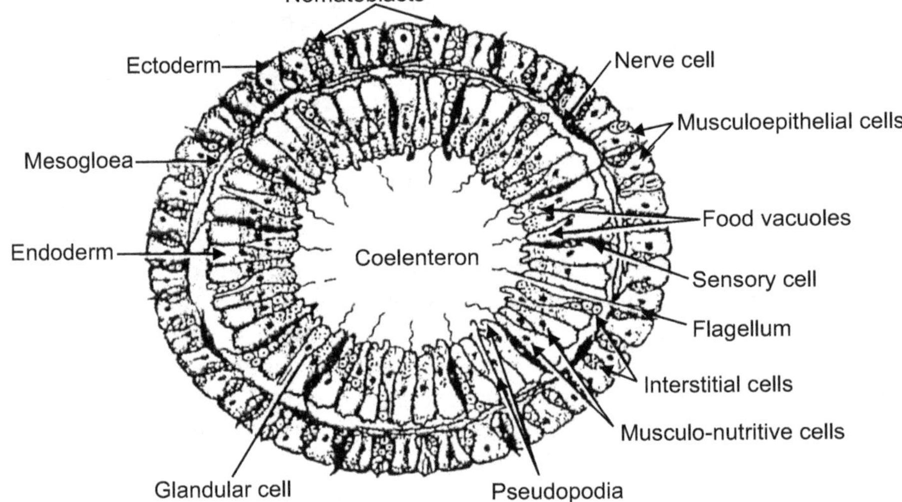

Fig. 1.24: *Hydra* T.S.

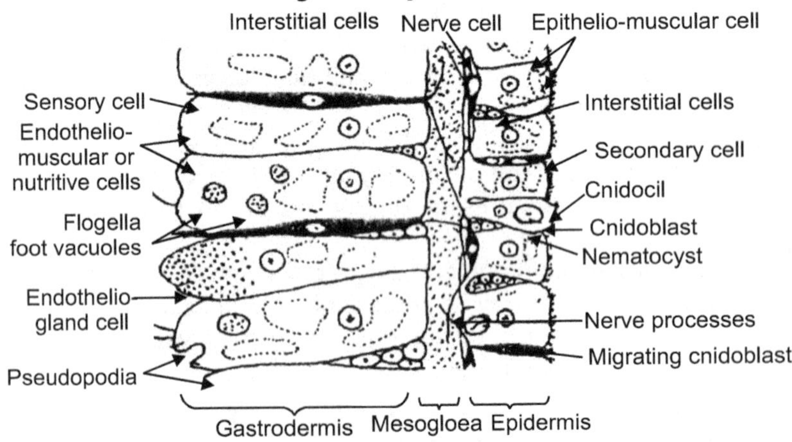

Fig. 1.25: *Hydra*. A portion of the section of body wall magnified

(i) The epitheliomuscular cells are cylindrical in shape with nucleus and supporting fibril. It has myonemes or unstriped muscle fibres at the base. There is row of granules which secrete the cuticle. The cells of basal disc are granular and they secrete mucus for attachment of *hydra*. These cells form protective covering and they help in contraction and locomotion.

(ii) The interstitial cells are lying in species between the inner ends of the cells of epidermis and between outer ends of cells of gastrodermis. They form nematocysts, germ cells, and epitheliomuscular cells. They also renew all cells of the animal once every 45 days. They also form gonads and rebuilding tissues during growth.

(iii) Guard cells are present at the mouth and pedal region.

(iv) Cnidoblast cells are found throughout the epidermis but specially on tentacles. They produce nematocysts.

(v) Sensory cells have large nucleus and from these cells sensory hair or flagellum is projecting out. They are found on both the germinal layers but more on ectoderm. They are sensory to touch, light, temperature changes and chemicals. They act as receptor and act both receives and transmits impulses.

(vi) The nerve cells and ganglion cells are small and elongated with one or more processes. They are situated at base of the epitheliomuscular cells. They are rarely found in gastrodermis.

(vii) Germ cells are formed by the repeated divisions of the interstitial cells. These form the gonads which later differentiated in either tests or ovaries. Gastrodermis is made up of large columnar epithelial cells with regular flat bases. Gastrodermis forms about two thirds of the body wall and is secretary, digestive, muscular and sensory. The gastrodermis has different cells like nutritive muscular, interstitial and gland cells. The cells are highly vacuolated and often filled with food vacuoles. The free ends of the cell usually bears two flagella. Nutritive muscular cells may also secrete digestive enzymes into the coelenteron for the digestion of food.

T. S. of *Hydra* passing through testis :

T. S. of hydra passing through the testis region, shows the following important features.

1. Testes are seen near the anterior end of the body and produced by the interstitial cells (germ cells) of the ectoderm and it encloses a cavity. Each testes is small and conical elevation, projects from the body surface.

2. Testes has outer covering of the ectoderm. Internally, the cavity contains a large number of spermatogonia spermatocytes spermatids and spermatozoa. Each spermatozoon has a oval head and a long tail.

3. At maturity, the testis reptures to release the spermatozoa.

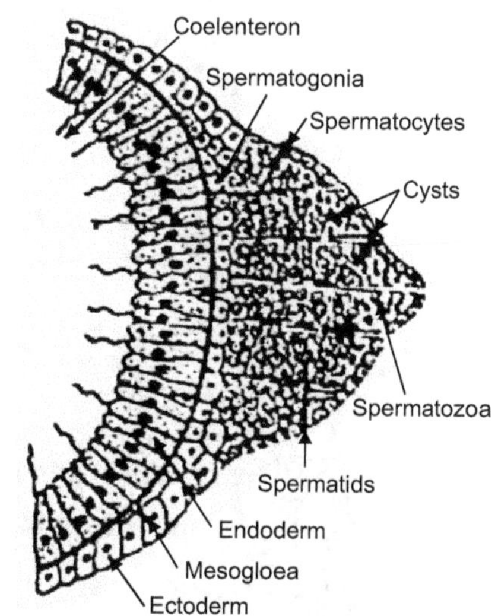

Fig. 1.26: *Hydra* T. S. through testis

4. The section also shows the diploblastic body wall containing ectoderm, mesogloea and endoderm and a central cavity, termed the coelenteron.

T. S. of *Hydra* passing through ovary :

T. S. passing through the region of ovary, shows the following important features.

1. Presence of one or more globular bulgings (outgrowths) near aboral end. Each bulding representing an ovary. It is easily recognised from the conical elevation of testis.
2. An ovary has an outer covering of ectoderm.
3. Each ovary encloses a single large female gamete or *ovum*. The mature ovum is a large spherical mass having a large nucleus and several yolk granules (food material).
4. At maturity the ovary wall ruptures and ovum remains ready for fertilization.
5. The section also shows two body layers, the *ectoderm* and *endoderm*, and structureless, gelatinuous *mesogloea*.

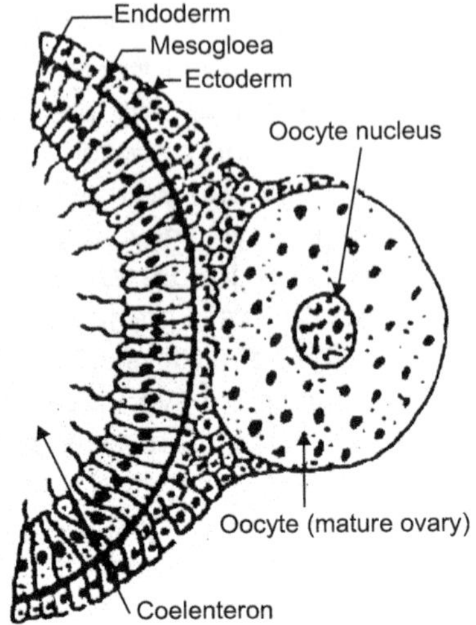

Fig. 1.27: *Hydra*. T. S. through ovary

5. TAENIA SOLIUM : SCOLEX AND PROGLOTTID

1. *Taenia solium* is the pork tapeworm of man as adult lives in the intestine of man leading an endoparasitic life.
2. Man is primary host and pig is secondary host hence it is called digenic. Other mammals like goat, cattle, horse and monkey may also serve as secondary host.
3. It lives in intestine of man, it causes injuries to mucous membrane lining of the alimentary canal where it adheres by its scolex.
4. It may cause mechanical injury by obstructing the passage of alimentary canal.
5. It causes abdominal pains, weakness, loss of weight and excessive appetite.

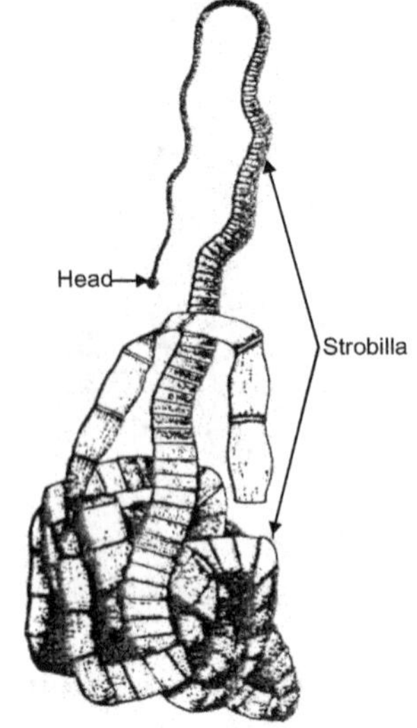

Fig. 1.28: An entire specimen

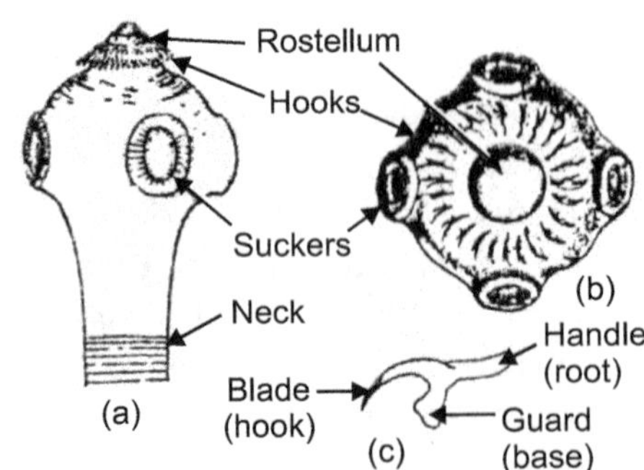

(a) Side view of scolex, (b) Frontal view of scolex, (c) Structure of a single hook

Fig. 1.29: Anterior end of *Taenia solium*

6. The disease caused by this worm is called *taeniasis*.
7. The body of tapeworm is long, dorsoventrally, flattened, narrow, ribbonlike, reaching a length of 2 to 3 metres.
8. Body consists of scolex, neck and strobila or body segments.
9. The scolex is know like, with four cuplike muscular suckers having radial muscles.
10. The anterior round prominence called rostellum has 22 to 32 curved, chitinous hooks in two circles.
11. The inner circle with larger hooks and outer circle with smaller ones.
12. Each hook is made of a *base* by which it fixes a *handle* directed towards the apex and a conical blade directed outwardly.
13. The scolex with its suckers and hooks is an organ of attachment to the intestinal wall of the host. Thus, it is called organ of adhesion.

Gravid Proglottid :

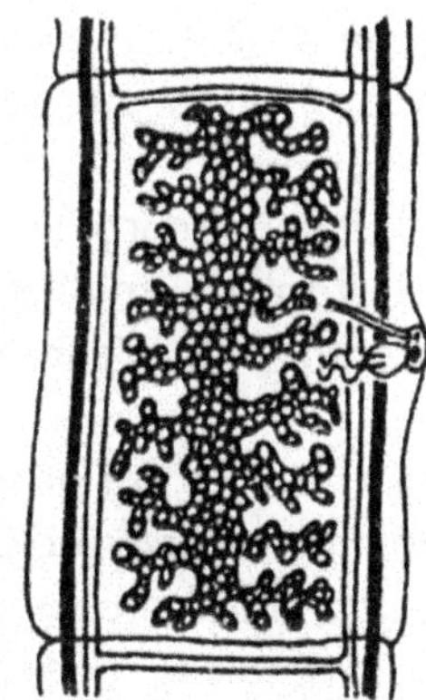

1. Gravid proglottids are the oldest and towards the posterior side of the strobila and include nearly 150 to 200 proglottids.

2. This segment is longer than its breadth and no reproductive organs are found in it.

3. It contains only branched uterus packed with fertilized eggs.

4. The gravid proglottids are regularly cut-off either singly or in group of 2 to 5.

Fig. 1.30: Gravid proglottis of *Taenia solium*

5. Process of cutting off is called apolysis.
6. These detached proglottids are passed out from the body of the host during defaecation along with the faeces. Then the embryos are transferred to secondary host like pig. Thus, the life cycle is continued.

✱✱✱

Practical **7**...

Aim : To identify any Three Museum Specimen with the Help of Taxonomic Identification Key.

1. ORDER - HEMIPTERA

Keys to Common Families of Hemiptera:

1. Hind leg without tarsal claws and having both tarsus and pretarsus flattened and bearing a dense fringe of long hair down each side, Fig. 1.32 A; middle tarsus having normal tarsal claws (part of **Heteroptera**). **2**

   Hind tarsus having tarsal claws similar to those on middle tarsus; hind leg usually without a long fringe but occasionally having one, Fig. 1.32 C. **3**

2. Beak forming a triangular striated piece that appears as a ventral sclerite of the head, Fig. 1.32 G; middle tarsus having extremely long claws, Fig. 1.32; front tarsus cornb like.

   Corixidae

   Beak cylindrical and rodlike, curving back from the ventral portion of the head, as in Fig. 1.32 L: front and middle legs in shape. **Notonectidae**

3. Beak arising from posterior margin of head. Fig. 1.31 L, no gula present behind it

   Homoptera

   Beak arising from front or venter of head, Fig. 132 L, the venter of the head posterior to beak forming a sclerotized bridge or gula. **Heteroptera 15**

4. Having wings, which are sometimes reduced to short scales. **5**

   Completely wingless species. **13**

5. Front femur greatly enlarged in comparison with middle femur, Fig. 1.31 K; three ocelli present **Cicadiae**

   Front femur no larger than middle femur; three, two or no ocelli present. **6**

6. Antennae arising from sides of head, situated beneath or behind eyes, Fig. 1.31 M.

   Fulgoridae

   Antennae arising from front of head between eyes, Fig. 1.31L. **7**

7. Pronotum enlarged dorsally into a large structure which covers most of head and body and may be highly ornamented with spines and processes, Fig. 1.34. e.g. **Membracidae**.

   Pronotum much smaller, without dorsal enlargement. **8**

8. Pronotum forming a broad shield which covers the greater part of the mesonotum, Fig. 1.37; tarsus 3-segmented, Fig. 1.31H, I. **9**

   Pronotum forming a narrow collar which does not extend back over the mesonotum, tarsus 1- or 2-segmented. **10**

9. Hind tibia bearing a double row of spines down its entire length, its apex usually not enlarged, Fig. 1.31H. e.g. **Cicadellidae**

   Hind tibia with only scattered spines except at apex, which is enlarged and armed with a prominent crown of spines, Fig. 1.31 I. e.g. **Cercopidae**

10. Having only one pair of wings. E.g. **male Coccoidea**

    Having two pairs of wings. **11**

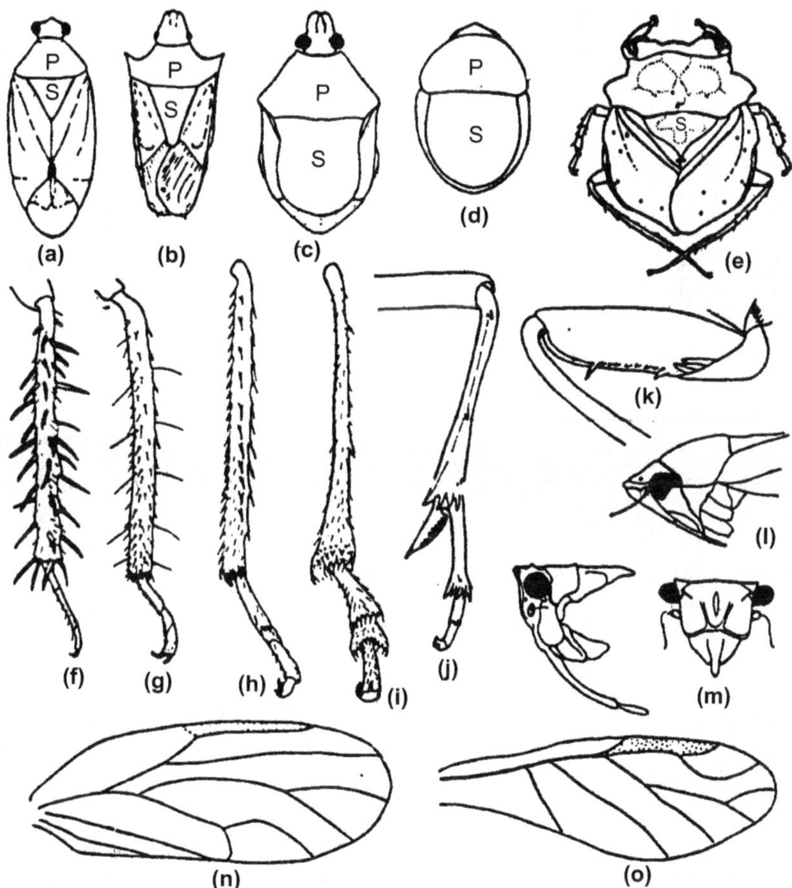

Fig. 1.31: Diagnostic characters of Hemiptera. A. Outline of *Lygus*. Miridae; B. Outline of *Solubia*, Pentatomidae; C. Outline of *Stiretrus*, Pentatomidae; D. Outline of *Corimelaena*, Pentatomidae; E. *Gelastororis*, Gelastocoridae; F. Hind tibia and tarsus of *Pangaeus*, Pentatomidae; G. Hind tibia and tarsus of *Thyanta*, Pentatomidae; H. Hind tibia and tarsus of *Aulacices*, Cicadellidae; I. Hind tibia and tarsus of *Aphrophora*, Cercopidae; J. Hind tibia and tarsus of *Stemocranus*. Fulgoridae; K. Front femur of *Pacarina*, Cicadidae; L. Head of *Gypona*. Cicadellidea; M. Head, lateral and anterior views, of *Publicia*, Fulgoridae; N. Wing of *Psylla*, Psyllidae; O. Wing of *Aphis*, Aphididae.

11. Wings milky-opaque, covered with a fine powdery white wax.   **Aleurodidae**

   Wings transparent or patterned, not covered with a waxy secretion. **12**

12. Front wing with R_s very long, arising before stigma, and Cu branched, Fig. 1.32; abdomen never with cornicles .... **Psyllidae**

Front wing with R_s short, arising from some part of the stigma, and Cu unbranched, Fig. 1.31 O; abdomen in many species having a pair of lateral tubes or cornicles.

Aphidoidea

13. Eyes large, antennae situated at sides of head below or behind eyes, Fig. 1.31 M.

Fulgoridae

Either eyes rudimentary or absent, or antennae situated on front of head between eyes, Fig. 1.39. .... **14**

14. Tarsus 1-segmented; body covered with a hard shell, waxy secretions, or a detachable scale, Fig. 1.34; abdomen never having cornicles.   **Coccoidea**

Tarsus 2-segmented; body at most with waxy secretions; abdomen often having a pair of conspicuous cornicles or tubes, Fig. 1.34.   **Aphidoidea**

15. Antennae shorter than head, usually recessed in a concavity beneath the eyes or under the lateral margin of the head, as in Fig. 1.32 G. **16**

Antennae at least as long as the head, usually extending free from it, Fig. 1.32, sometimes fitting into a pronotal groove when at rest. **18**

16. Ocelli present. Small toadlike bugs, Fig. 1.31E, found along the margins of lakes and streams.   **Gelastocoridae**

Ocelli absent. Forms living in water, sometimes flying and attracted to lights. **17**

17. Tarsi 1-segmented, front tarsus with only a minute claw or none, Fig. 1.32D; apex of abdomen with a long or, short respiratory tube, Fig. 1.32 B, each blade of which is concave mesally, the two fitting together to make a hollow tube: hind legs slender and without fringes of long hair.   **Nepidae**

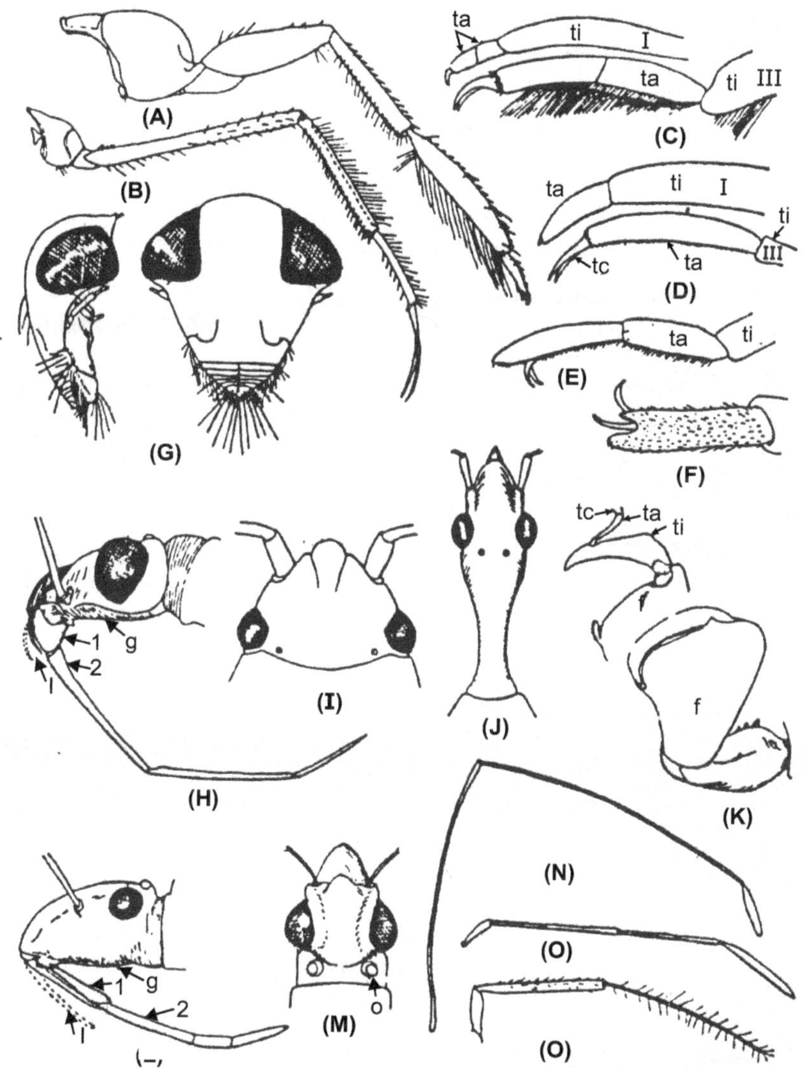

Fig. 1.32: Diagnostic characters of Hemiptera. A, and B, hind leg and middle leg of *Corixa*, Corixidae; C, front and hind tarsi of *Belostoma*, Belostomatidae; D, front and hind tarsi of *Nepa*, Nepidae; E and F, Front tarsus, Lateral and dorsal aspect of cerris, Cervidae G. head of Corixa; H, head of *Nebis*, Nabidae; I, head of *Lygaeus*, Lygaeidae; J, head or Myodocha, Lygaeidae; K, front leg, insect showing structure of tarsus of *Phymata*, Phymatidae, L, head of Alydus, Coreidae; M and N, head and antenna of *Jalysus*, Neididae; O, antenna of *Myodocha*, Lygaeidae; P, antenna of *Lyctocoris*, Anthoxoridae, *f*, femur; g. gula; *l*, labrium; o, ocellus; ta, tarus; tc, tarsal claws; ti, tibia

Tarsi 2-segmented, front tarsus with a stout curved claw, fig. 1.32C; apex of abdomen at most with a pair of short flat respiratory filaments; hind tibia and tarsus often flattened, always having fringes of long hair for swimming   **Belostomatidae**

18. Head extremely long and slender, slightly bulbous at apex, where beak arises, the eyes situated at the middle of what appears to be a long neck; rest of body also very slender, Fig. 1.32C.   **Hydrometridae**

    Head much stouter, Fig. 1.32I, or eyes not situated on the neck, Fig. 1.32J. **19**

19. Front leg having femur and tibia chelate, forming a large grasping device, femur swollen and triangular, tibia curved and closing against the end of femur, Fig. 1.32K **Phymatidae**

    Front leg not chelate. **20**

20. Claws of front tarsus inserted before apex, Fig. 1.32E, F. **21**

    Claws of front tarsus attached at apex, as in Fig. 1.32C. **22**

21. Middle pair of legs attached far from front legs, close to hind legs; hind femur very long, Fig. 1.32, beak 4-segmented   **Gerridae**

    Middle pair of legs attached about midway between front and Hind legs; hind femur only moderately long; beak 3-segmented.   **Veliidae**

22. Scutellum very large, reaching about one-half or more distance from posterior margin of pronoturn to end of folded wings, Fig. 1.31B-D; antennae usually 5-segmented,

                                                                                 Pentatomidae

    .... **23**

    Scutellum much smaller, reaching about a quarter of the distance from pronoturn to tip of body, Fig. 1.31A; antennae usually 4-segmented. **26**

23. Tibia armed with rows of thick thornlike spines, Fig. 1.31F **24**

    Tibia having series of short even spines, occasionally with a few scattered, very slender hairs, Fig. 1.31G. **25**

24. Scutellum triangular and not very large, as in Fig. 1.31B   **Cydninae**

    Scutellum large and U-shaped, covering most of abdomen, Fig. 1.31D   **Thyreocorinae**

25. Scutellum U-shaped and very wide, covering almost all of abdomen, the sides of scutellum curved mesad at extreme base, as in Fig. 1.31D   **Scutellerinae**

    Scutellum V-shaped, Fig. 1.31B, or, if U-shaped, then never larger than in Fig. 1.31C, and slightly contracted just beyond base   **Pentatominae**

26. Front wing abbreviated, with no membrane, and reaching at most to middle of abdomen, Fig. 1.33. .... **27**

    Front wing normal, with a large apical membrane, or reaching well beyond middle of abdomen, Fig. 1.33. .... **28**

27. Body flat and wide; front wing short, broad, and scalelike, only barely reaching over base of abdomen; sides of pronoturn large, round, and flange like, Fig. 1.33 beak 3-segmented; antennae long and slender   **Cimicidae**

Body narrower, or otherwise different from foregoing, having either a 4-segmented beak, different-shaped wing, or short antennae. A few genera, most of them rare, difficult to key to family, belonging to **Anthocoridae, Miridae, Aradidae, Lygaeidae**, or **Nabidae**; and all nymphs of Heteroptera families listed beyond this point. The wingless species of this group are keyed no farther here.

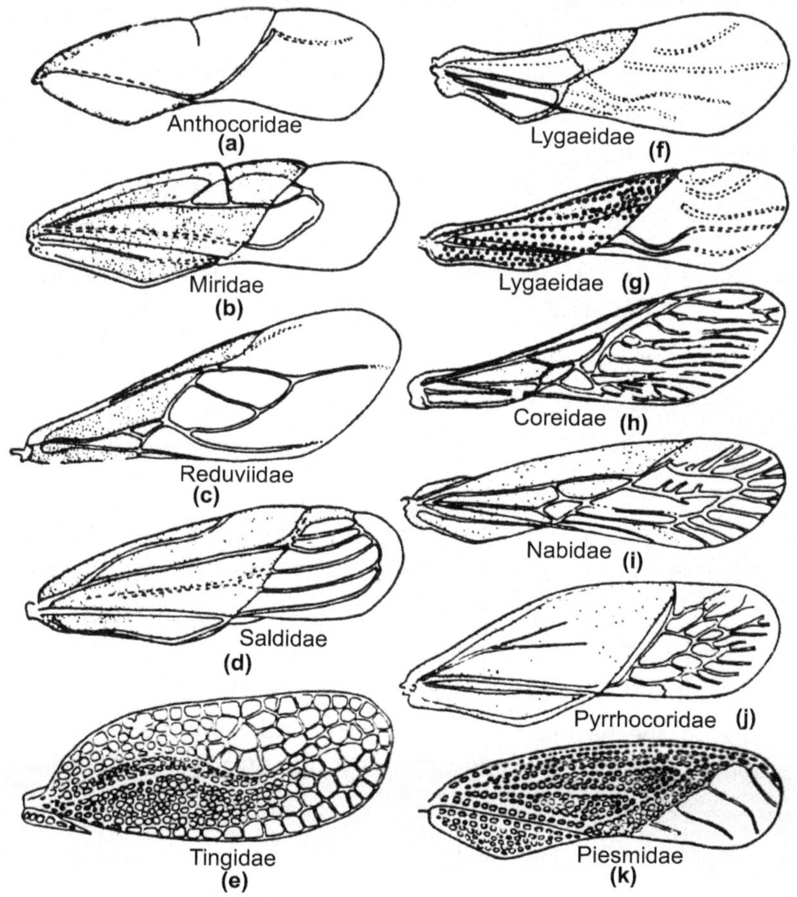

Fig. 1.33: Elytra or forewings of Hemiptera. A, *Triphleps*; B, *Lygus*; C, *Pselliopus*; D, *Salda*; E, *Gargaphia*; F, *Blissus*; G, *Myodocha*; H, *Alydus*; I, *Nabis*; J, *Euryophthalmus*; K, *Piesma*.

28. Hemelytra large, covering entire abdomen and reticulate over their entire surface with a net-like pattern, with little or no distinction between corium and membrane, Fig. 1.33E   **Tingidae**

Hemelytra with a definite apical membrane, Fig. 1.33, all except E............................ **29**

29. Membrane of hemelytron having one or two large basal cells, and none, one, or two short spurlike veins extending distally from these, Fig. 1.33B, C. **30**

    Membrane of hemelytron having either no closed cells, Fig. 1.33A, or at least five or six veins (including the costa) running through the membrane, Fig. 1.33D, H, I, J. **31**

30. Ocelli prominent, two in number, membrane of hemelytron having a long vein proceeding from top of upper closed cell, Fig. 1.33C. 　　　　　**Reduviidae**

    Ocelli absent; membrane with a vein proceeding only from bottom of lower closed cell, or such a vein lacking, Fig. 1.33B. 　　　　　**Miridae**

31. Antenna having first two segments stout, last two threadlike, forming a slender terminal filament, Fig. 1.32P; ocelli present but small; hemelytral membrane with only one or two weak veins, Fig. Fig. 1.33A 　　　　　**Anthocoridae**

    Antenna having one or both of the two apical segments as thick as the first or second, Fig. 1.33N, O. **32**

32. Hemelytral corium extending markedly beyond a ridge like oblique vein near apex of corium, Fig. 1.33K; corium entirely reticulate 　　　　　**Piesmidae**

    Hemelytral corium not extending beyond an apical oblique vein, Fig. 1.33H, J.

    or not having such a vein, Fig. 1.33G **33**

33. No ocelli present **34**

    Two ocelli present. **35**

34. Flat wide wart bugs, Fig. 1.32 A, B; tarsus 2-segmented, the first segment short; hemelytra often small, the periphery of the abdomen extending considerably beyond them. 　　　　　**Aradidae**

    Stout insects, the body deep; tarsus 3-segmented, the first segment long; hemelytra larger, Fig. 1.33J, covering all abdomen except tip and sides near apex 　　**Pyrrhocoridae**

35. Hemelytral membrane having four or five large and fairly regular closed cells and no other venation, Fig. 1.33D; oval fairly flat bugs found on stream and lake shores. **Saldidae**

    Hemelytral membrane having either an irregular network of cells or only one or two small, well-sclerotized ones, Fig. 1.33F-I **36**

    Membrane having a series of about 15 irregular veins, at least on apical portion,

    Fig. 1.33H, I. **37**

    Membrane having only five or six veins across it, Fig. 1.33F, G.......................... **38**

37. First segment of beak short and conelike, thicker than the second, Fig. 1.32H; front femur thickened, front tibia armed inside with a double row of short black teeth 　　**Nabidae**

First segment of beak cylindrical and long, similar in general shape to second segment, Fig. 1.32L; front femur usually more slender, front tibia never with inner rows of black teeth

Coreidae

38. Each ocellus situated behind an eye, at the base of a distinct swelling, Fig. 1.32M; extremely slender and elongate bugs, with long and slender legs and antennae; last segment of antenna short and oval, forming a small club, Fig. 1.32N   **Neididae**

39. Ocelli situated closer to or between eyes, and not at the base of a swelling, Fig. 1.32I, J; chiefly robust short insects or having short legs; antennae either short, Fig. 1.32O, or not clubbed

Lygaeidae

Suborder Homoptera:

Cicadas, leafhoppers, aphids, scale insects, etc. This suborder contains the cicadas, leafhoppers, aphids, scale insects, and their allies, all plant feeders. The North American fauna is composed of about a dozen families or superfamilies.

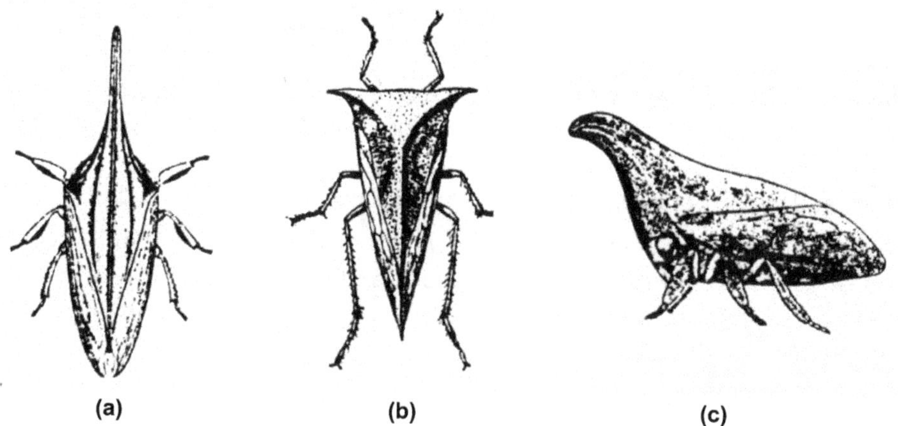

(a)   (b)   (c)

Fig. 1.34: Treeehoppers, Membracidae. A, Campylenchia latipes; B, Ceresa bubalis; C. Enchenopa binolata (A, C, from Kansas Stage College; B from U.S.D.A.)

Fig. 1.35: A fulgorid *Peregrinus maidis* (After Thoma)

The Fulgorid-Cicada Series : In this series, the flagellum of the antenna is needle-like and the tarsus is almost invariably 3-segmented. In the series are found many forms of bizarre appearance. Some of the Membracidae, or treehoppers, Fig. 1.34, have the pronotum greatly enlarged and ornamented with ridges, horns, or prongs. The Fulgoridae are a large family, and many resemble leafhoppers, Fig. 1.34. Some of the fulgorids have large foliaceous wings, and others, such as our native *Scolops* and the South American lantern fly, or peanut bug, *Lantemaria phosphorea*, have bizarre projections of the head. Another oddity is the spittle bug family, Cercopidae. The nymphs of this family produce masses of white froth or spittle-like substance and live hidden beneath it. Two well-known and abundant families of the series are the Cicadidae (cicadas) and the Cicadellidae (leafhoppers).

Cicadidae, Cicadas : These are large insects, many North American species measuring 2 inches or more. They are distinguished structurally from related families by having three distinct ocelli on the dorsum of the head. The males have highly developed musical organs, and during warm days and summer evenings they make a shrill noise.

2. ORDER : COLEOPTERA

Key To Common Families of Coleoptera :

1. Front of head produced into a definite beak, that may be long or short, the antennae arising from side of beak; palpi vestigial.　　　　　　**Curculionidae**

   Front of head not produced into a beak; if slightly so, antennae arising between the eyes, or maxillary palpi prominent, Fig. 1.36..**2**

2. Middle and hind legs very wide and flat, almost paper-thin, fitted for swimming, the basitarsus large and triangular, the next two produced laterally to form long swimming "fingers"; front legs tubular, fitted for grasping, Fig. 1.36L; each eye completely divided, one part on dorsum of head, the other on ventral aspect of head, Fig. 1.36K.　　**Gyrinidae**

   Middle and hind legs having some or most segments robust and not flattened, occasionally furnished with rows of long hairs and fitted for swimming in this fashion; eye seldom divided, and then only by a continuation of a head flange or by an antennal base, Fig. 1.36K. **3**

3. Maxillary palpi longer than antennae and slender, resembling antennae, chiefly aquatic.　　　　　　**Hydrophilidae**

   Maxillary palpi shorter than antennae, and not antenna-like; antennae various.　.........**4**

4. Antenna with each of last 3 to 7 segments enlarged on one side to form an eccentric plate or lamella, Fig. 1.36A, C, each lamella situated nearly at a right angle to long axis of antenna. **5**

   Antenna not lamellate, but frequently having the end segments fairly evenly enlarged to form a club, Fig. 1.36E, or most of the segments with anterior projections giving a saw tooth. Fig. 1.36G, J, or pectinate outline. **6**

5. Elytra short and squarely truncate, exposing three full tergites of abdomen, these tergites heavily sclerotized and hard, Fig. 1.36B.　　　　　　**Silphidae**

   Elytra longer, usually rounded at apex, usually covering entire dorsum of abdomen but in a few species exposing one or two tergites, Fig. 1.36.　　　　**Scarabeidae**

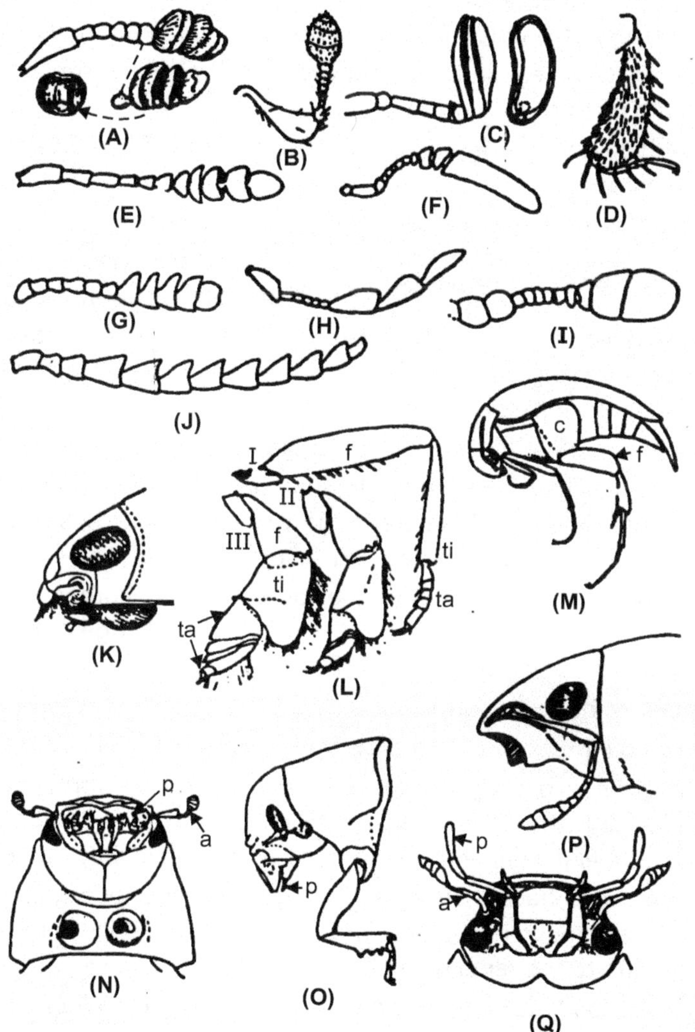

Fig. 1.36: Diagnostic characters of Coleoptera. A, antenna of *Nicrophorus*, Silphidae, insects showing concave end segments and an end view of one, B, antenna of *Scolytus*, Scolytidae; C, antenna of *Thyce*, Scarabeidae; D, tibia and tarsus of *Heterocerus*, Heteroceridae; E, antenna of *Silpha*, Silphidae; F, antenna of *Attagenus*, Dermestidae; G, antenna of *Languria*, Languriidae; H, antenna of anobiidae; I, antenna of *Anthrenus*, Dermestidae; J, antenna of *Melanotus*, Elateridae; K, head of *Dineutes*, Gyrinidae; L, legs of Dineutes, Gyrinide; M, profile of *Mordellistena*, Mordelliadae; N and O, head and prothorax of Dendroclonus, Scolytidae, P, head of a short-snouted weevil, curculivnidae, Q, head of *Tropisternus*, Hydrophilidae, antenna; c, coxal plate, f, femur; p. maxillary palpus; ta, tarsus, ti, tibia.

6. Elytra short, exposing five or more sclerotized abdominal tergites, Fig. 1.36A **7**
   Elytra covering all or most of Abdomen, never more than two or three sclerotized tergites visible from above; occasionally the abdomen of a female of this group may be

extremely distended with eggs, and a third or fourth segment may project from beneath the elytra, but all except the apical two are soft and semimembranous. **8**

7. Elytra truncate, parallel-sided, and abutting evenly down the meson, Fig. 1.36; abdomen hard and regular in outline **Staphylinidae**

   Elytra ovate, overlapping considerably at base; abdomen flabby and shrinking irregularly when the specimen dries **Melloidae**

8. Hind tarsus 4-segmented, front and usually middle tarsi 5-segmented. **9**

   Either hind tarsus having 3 or 5 segments, or all tarsi having the same number of segments. **11**

9. Body narrow and deep, bilaterally compressed, and with the hind coxa forming a large plate that appears as a major sclerite in the side of the thorax, Fig. 1.36M; small beetles often found abundantly in flowers. **Mordellidae**

   Body wider than deep, often flattened, hind coxae no larger than in Fig. 1.36. .... **10**

10. Each front coxal cavity closed posteriorly by a projection of the pleuron which meets the apex of the sternum, Fig. 1.36 D; lateral edge of pronotum. forming a sharp flange or delineated by a ridge or carina (compare Fig. 1.36H) **Tenebrienidae**

    Front coxal cavities open posteriorly, the posteromesal corner of propleuron not extending mesad of outer portion of coxa, Fig. 1.36C; lateral edge of pronotum rounding inconspicuously into pleural region. **Meloldae**

11. Head not retracted within prothorrax. **12**

    Head retracted within prothorax, so that only anterior portion protrudes, Fig. 1.36. ... **14**

12. Ventral portion of head having a large convex gular region with a single gular suture down the middle; and the palpi very short or indistinct, Fig. 1.36N; antennae always elbowed, with a long first segment, and sometimes ending in a flat club, Fig. 1.36B. ... **13**

    Ventral portion of head having either a small gular area or two gular sutures, and the maxillary palpi usually much longer Fig. 1.36; antennae not elbowed or the first segment shorter in proportion to the remainder. **14**

13. Antenna long, Fig. 1.36P, usually ending in a small cylindrical, elliptic club; side of head having a deep groove for reception of antenna. **Curculionidae**

    Antenna short, Fig. 1.36N, O, ending in a large flat club or comb, Fig. 1.36B; side of head without antennal groove **Scolytidae**

14. Hind and middle tarsi either 3- or 4-segmented, or fourth segment very small in comparison with the third, Fig. 1.37b, c. **15**

    Hind and middle tarsi 5-segmented, the fourth as large as or as thick as the third, Fig. 1.37a. **27**

15. All tarsi having the third segment enlarged, Fig. 1.37b, c, deeply bilobed or channeled dorsally; the fourth segment extremely small, either sunken into the cleft of the third or arising from dorsum of the base of the third, Fig. 1.37b; fifth segment large and normal; frequently the fourth segment cannot be seen, or it appears as a minute subdivision of the base of the fifth segment, in which case each tarsus appears 4-segmented. **16**

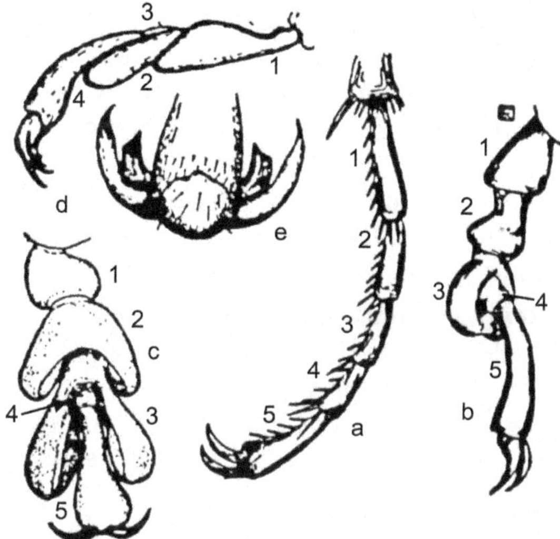

Fig. 1.37: Tarsi of Coleoptera. a. *Harpalus*, Carrabidae; b. Leptinotarsa, Chrysomelidae; c. *Chelymorpha*; d. *Epilachna*, Coccinellidae; e. toothed tarsal claws (From Matheson, Entamology for Introductory Courses, by Permission of the Comstock Publishing (Co.)

Either tarsi only 3-segmented, or third segment not enlarged, channeled, or bilobed....**42**

16. Antennae about as long as or longer than the body, Fig. 13.6   **Many Cerambycidae**
    Antennae distinctly shorter than body.. **17**

17. Last tergite (pygidium) exposed, almost completely visible beyond or below end of elytra, Fig. 1.36E, F. **18**
    Lattergite almost entirely or completely covered by elytra. **21**

18. Hind tibia little, if any, longer than basitarsus, and having a pair of long apical spurs, Fig. 1.36P   **Bruchidae**
    Hind tibia much longer than basitarsus, frequently with only short spurs or none. **19**

19. Elytra short, each only about twice as long as wide; stocky short species, the pygidium oblique, and at a definite angle to dorsal contour of the body, Fig. 1.36 F. **20**
    Elytra long, each four times or more as long as wide; elongate species, the pygidium nearly horizontal, following dorsal contour of the body, Fig. 1.36 ... some **Cerambycidae**

20. Anterior part of prothorax forming a cylinder against which the flat head fits like a lid; eyes oval, fitting against margin of prothorax, Fig. 1.36 H .... some **Chrysomelidae**
    Anterior part of prothorax narrow, head projecting freely beyond it; eyes incised and V-shaped, head constricted behind them, Fig. 1.36   most **Bruchidae**

21. Last 3 to 5 antennal segments enlarged to form a large loose club, Fig. 1.36G; elongate, smooth, and highly polished species.   **Languriidae**
    Antenna of uniform thickness throughout, or widening gradually to form a club, or forming a round compact club; form various. **22**

22. Hind femur greatly enlarged, short, and oval in outline; flea beetles and their relatives
                                                                                          Chrysomelidae

Hind femur elongate or more parallel-sided, or constricted at base and enlarged only at apex, Fig. 1.36L. **23**

23. First abdominal sternite very long, its mesa length nearly equal to that of the following four sternites, combined, Fig. 1.36 G; several genera of varied form (e.g., Donacia, Crioceris)   **Chrysomelidae**
First abdominal sternite considerably shorter in proportion to following segments ... **24**

24. Either antenna as long as or longer than body, or hind femur having basal portion slender and apical portion enlarged and clavate, Fig. 1.36   **Cerambycidae**
Antenna shorter than body, and hind femur without a basal stalk-like portion. **25**

25. Tibia having a pair of well-developed tibial spurs at apex   Cerambycidae
Tibia having either no tibial spurs or very minute ones.

26. Mesal margin of eye deeply incised, with the antenna situated in the incision, Fig. 1.36K, or the eye completely divided, the antenna situated between the two parts.
                                                                                        Cerambycidae
Either eye not incised, or antenna not at all in incision of eye.   **Chrysomelidae**

27. Hind coxa having a wide long ventral plate which covers coxa, trochanter, and most of the femur, Fig. 1.36A; small to medium-sized stout aquatic beetles.   **Haliplidae**
Hind coxa at most having only a small outer plate, which does not cover trochanter or femur. **28**

28. Apices of hind coxae forming a double-knobbed process; base of each coxa extremely large and platelike, appearing as a dominant sclerite of the sternum, hind legs fringed for swimming.   **Dytiscidae**
Hind coxa neither with such a knobbed apex nor with such a platelike base. **29**

29. First abdominal sternite completely divided by the hind coxal cavities, the sternite appearing as a pair of triangular sclerites, one on each side of the coxae; first three abdominal sternites immovably united, the separating sutures appearing partly as extremely fine lines. **30**
Either first abominal sternite not completely or not at all cut into by the coxal cavities, or the first three sternites with separating sutures extremely well developed across the entire width of the segment, as are those between the more apical segments, Fig. 1.31

30. Clypeus fairly narrow, the antennal sockets wider apart than the width of the clypeus, Fig. 1.36N   **Carabidae**
Clypeus much wider, Fig. 1.36M, the antennal sockets situated closer together than the width of the clypeus.   **Cicindelidae**

31. Abdomen having 6 or more exposed sternites. **32**
Abdomen having not more than 5 exposed sternites. **35**

32. Last 3 to 5 antennal segments enlarged to form a club, Fig. 1.36E   **Silphidae**
Antenna the same width throughout, often beadlike or serrate............................. **33**

33. Tarsus slender and smooth, segments 1 to 4 very short and ringlike, segment 5 long, about equal in length to first 4 combined, aquatic forms   **Psephenidae**
Tarsus stout and densely setose, segment 5 no longer than segment 1, some of the basal segments elongate... **34**

34. Prothorax with broad anterior and lateral margins forming a hood that covers most or all the head from above, head partially retracted, viewed from side. **Lampyridae**
Prothorax not having wide margins, head not retracted in pronotum, hanging down or projecting forward freely **Cantharidae**

35. Antenna elbowed and capitate, the club appearing as one round segment, sometimes having faint cross sutures; very hard shining black beetles having short stout legs, the tibiae expanded and spurred for digging. **Histeridae**
Antenna not elbowed, and either not thickened toward tip, or the enlarged portion composed of 2 or 3 well-separated segments or a single elongate segment; legs or body shape different from above. **36**

36. Antennae elongate and serrate throughout, Fig. 1.36J.. **37**
Antennae short, filiform, or clavate. **38**

37. Pronotum having a sharp projection at each posterolateral corner. **Elateridae**
Pronotum having posterolateral corners rounded and not pointed. **Buprestidae**

38. Prosternum produced anteriorly to form a long concave shelf under the head, but the latter largely retracted into the opening of the prothorax. **39**
Prosternum not so produced, much shorter than pronotum and proportioned more, head usually held downward against chest. **40**

39. Front coxae round, situated some distance from lateral edge of pronotum; antennae elongate or definitely clavate. **Elmidae**
Base of front coxae triangular, the point of the triangle extending nearly to lateral edge of prosternum, antennae short, the flagellum of equal thickness throughout, with wide flat segments. **Dryopidae**

40. Antennae ending in a distinct club composed of 1 to 3 segments, Fig. 1.36F, I **Dermestidae**
Last 3 segments of antenna greatly enlarged but well separated to forma chain, Fig. 1.36H. **41**

41. Tibiae without spurs; small convex forms including drugstore and cigarette beetles, infesting dried food products. **Anebiidae**
Tibiae with distinct spurs, Fig. 1.36O; includes many subcylindrical forms ornamented with horny processes and ridged areas; the powder post beetles. **Bostrichidae**

42. Front and middle tarsi 4-segmented, hind tarsus 3-segmented. **Cucujidae**
All tarsi having the same number of segments. **43**

43. Tarsi 4-segmented, all segments well marked, and second and third of about equal width. **44**
Tarsi 3-segmented, or third segment minute and hidden at the base of an enlarged second segment, Fig. 1.37d. **45**

44. Tibiae dilated and armed with a series of stout spurs, fitted for digging, Fig. 1.36D; molelike beetles covered with dense pile, inhabiting wet mud or sand banks. **Heteroceridae**
Tibiae either not dilated or not armed with a series of spurs, flat beetles found under bark or in stored grain and feed **Cucujidac**

45. Second segment of tarsus dilated, with a large ventral pad, Fig. 1.37d; almost hemispherical beetles, usually polished and strikingly patterned.   **Coccinellidae**

Second segment of tarsus not much if at all wider than first segment; beetles often somewhat angular in outline, and flat; frequenting flowers, and tree sap.   **Nitidulidae**

3. ORDER - LEPIDOPTERA

Key to Common Families of Lepidoptera :

1. Wings reduced to small pads or entirely lacking, Fig. 1.38. .... **1**

    Wings well developed, atleast nearly as long as abdomen, Fig. 1.38. .... **2**

2. Legs lacking or reduced to short stubs; usually associated with a baglike case. **Psychidae**

    Legs elongate and normal in appearance. **3**

3. Abdomen having closely set scales or spines, or bristling dark-gray hair; usually not found near cocoon   **Geometridae**

    Abdomen smoothly clothed with fine light woolly hair; moth usually found clinging to cocoon.   **Liparidae**

4. Front wing having a jugum, a lobe at the base of the posterior margin for use in wing coupling, front and hind wings similar in venation and shape (jugatae).   **Hepialidae**

    Front wing without a jugum; either anterior margin of hind wing having an enlarged lobe at base, Fig. 1.39D, H, I, or with a long basal spine, or *frenulum*, Fig. 1.39E, both used for wing coupling; or front and hind wings markedly different in shape and venation. ...**5**

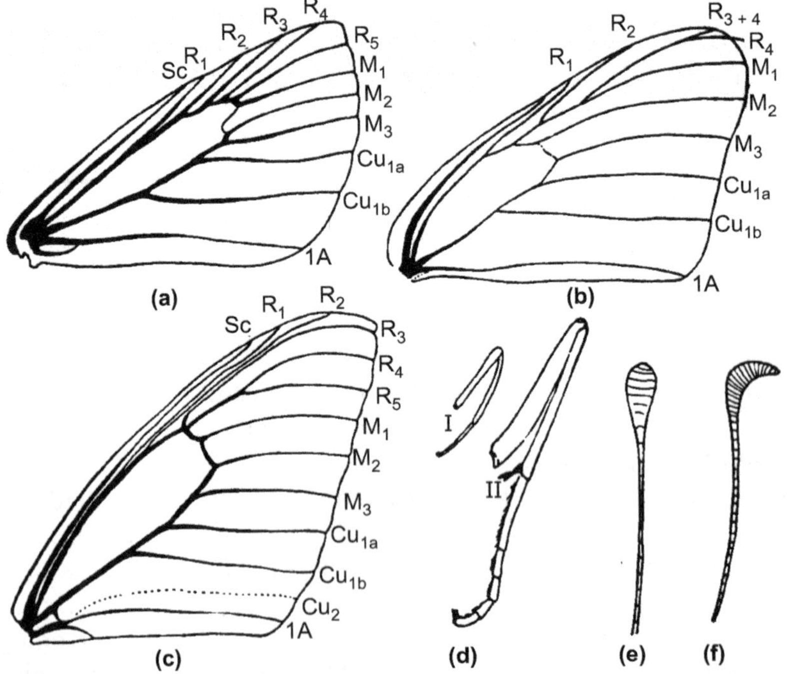

Fig. 1.38: Diagnostic parts of Lepidoptera. A, front wing of *Pamphila*, Hesperiidae; B, front wing of *Pieris*, Pieridae; C, front wing of *Papilio*, Papilionidae; D, front (small) and middle legs of *Brenthis*, Nymphalidae; E, tip of antenna of same; F, tip of antenna of *Thanaos*, Hesperiidae.

5. Hind wing without a frenulum, and antenna clubbed or hooked at apex, Fig. 1.38E, F.
    Rhopalocera

    Either hind wing having a Frenulum, or antenna not clubbed or hooked; instead either threadlike, or serrate, or pectinate   **Frenatae** **11**

6. Front wing having each of the 5 branches of radius and 3 of media arising from the discal cell, Fig. 1.38A; antennae usually hooked at apex, Fig. 1.38F   **Hesperiidae**

    Front wing having some of these fused at base, branching beyond discal cell, Fig. 1.38B, C. **7**

7. Front wing having Cu_1 appearing 4-branched, Fig. 1.38C, because both M_2 and M_3 are more closely associated with it than with R_5.   **Papilionidae**

    Front wing having Cu_1 appearing 3-branched, Fig. 1.38B, because M_2 is more closely associated with R_5. .... **8**

8. Labial palps longer than thorax, thickly hairy, and extending forward.   **Libytheidae**

    Labial palps shorter than thorax. **9**

9. Front legs reduced, and much shorter than the other legs, Fig. 1.38D   **Nymphalidae**

    Front legs larger in proportion to the others. **10**

10. Front wing having M_1 fused for a considerable distance with posterior branch of radius, Fig. 1.38B; colours white, yellow, or orange, plus black marks.   **Pieridae**

    Front wing having M_1 either not fused with R, or only slightly so, thus arising from discal cell or very near it; colours coppery, blue, or brown.   **Lycaenidae**

11. Hind wing having a posterior fringe as long as wing is wide, Figs. 1.40, wing usually lanceolate. A large number of families of small moths difficult to identify, including Yponomeutidae, Gelechiidae, and Tineidae   Not keyed further here

    Hind wing markedly wider than its fringe. **12**

    Front wing narrow, more than four times as long as wide; hind wing and sometimes front wing having transparent areas devoid of scales.   **Aegeriidae**

    Front wing wider, hind wings usually entirely covered with scales. **13**

13. Hind wing having 3 veins posterior to Cu_{1b}. **14**

    Hind wing having 1 or 2 veins posterior to Cu_{1b}, Fig. 1.39E. **15**

14. Front wing with Cu_2 fairly well developed, at least towards apex.   **Tortricidae**

    Front wing with Cu_2 atrophied.   **Pyralidae**

15. Front wing with 1A evenly bowed anteriorly, the vein coming close to central portion of Cu_1 or apex of Cu_{1b}, Fig. 1.39F.   **Sphingidae**

    Front wing with 1.4 straight or only slightly sinuate, Fig. 1.39C. **16**

16. Front wing having both M_2 and M_3 associated closely with Cu_1, which therefore appears 4-branched, Fig. 1.39C. **17**

    Front wing having M_2 either midway between M_3 and R_5, or closer to R_5, so that Cu_1 appears 3-branched, Fig. 1.39A. **20**

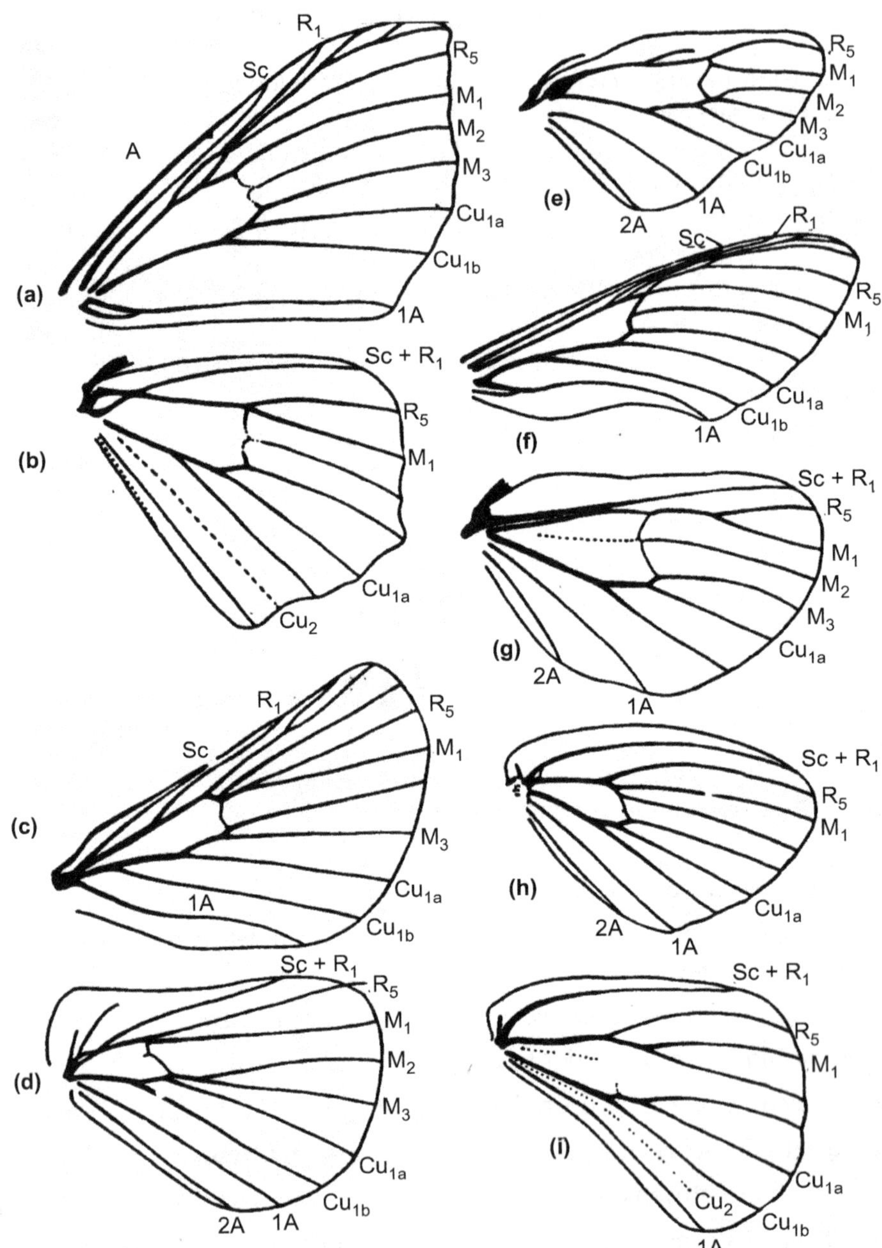

Fig. 1.39: Wings of Lepidoptera. A, B, front and hind, *Acidalia*, Geometridae; C, D, front and hind, *Malacosoma*, Lasiocampidae; E, hind, *Halisidota*, Arctiidata; F, front, *Protoparce*, Sphingidae; G, hind, *Hyperaeschra*, Notodonitadae; H, hind, *Citheronia*, Citheroniidae; I, hind, *Samia*, Saturniidae

Fig. 1.40: The tapestry moth *Trichophaga tapetzella*, one of the Microlepidoptera.

(From U. S. D. A., E.R.B.)

17. Hind wing without a frenulum, the base of the front margin greatly expanded, Fig. 1.39D

    Lasiocampidae

    Hind wing either with a frenulum or base of wing not expanded. **18**

18. Hind wing having Sc + P_1 and R_s, fused for about half length of distal cell, then the two veins separating, Fig. Fig. 1.39E   **Arctiidae**

    Hind wing having Sc + R_1, fused with R_s, for only a short distance, or the veins not at all fused. **19**

19. Head having two ocelli.   **Noctuidae**

    Head without ocelli.   **Liparidae**

20. Hind wing having Sc + R_1 arcuate, curving forward from its base and well separated from R_s, for its entire distance, Fig. 1.39H, I; frenulum obsolete. **21**

    Hind wing having Sc + R_1 either fused for a distance with R_s Fig. 1.39B, or running close to it, Fig. 1.39G; frenulum present, often tuftlike. **22**

21. Hind wing having two anal veins, Fig. 1.39H   **Citheroniidae**

    Hind wing having only one anal vein, Fig. 1.39I.   **Saturniidae**

22. Hind wing with Sc making a short sharp angulation at the base of the wing, Fig. 1.39B.

    Geometridae

    Hind wing with Sc not angulate at base, Fig. 1.39G   **Notodontidae**

∗∗∗

Practical **8**...

Aim : Visit to National Park.

For zoology students visit to zoological survey of India or Museum or National park is essential for the basic knowledge of the subject. You can get ample information and knowledge by arranging the visits to such places. Here, we will consider visit to National Park. Note the following points while arranging visit.

1. Fix the place of national park which is suitable and convenient to the student.

2. Before visiting the place, collect the information of that particular national park. Location, distance from your place, road map or railway route.

3. Which animals, plants are there, which is the best season to visit etc. information is available on net as well as in the books.

4. During the visit take all necessary permissions.

5. Strictly follow the rules and regulations of the national park.

6. Do not disturb the animals by making noise or feeding.

7. Carry a good camera/cell phone for photography.

8. Make the record of the animals and plants in your notebooks. Try to find out their scientific names.

9. Note the speciality of that national park.

10. Do not go very close to wild animals.

11. After your visit make the record of various animals, birds, plants which you saw.

12. Prepare your visit report along with good photographs.

***

Practical **1**...

Estimation of Dissolved Oxygen from given Water Sample

Aim : To estimate percentage of dissolved oxygen from given water sample.

Introduction :

Dissolved oxygen (D.O.) is a measure of the ability of a water body to support a well-balanced aquatic life. Insufficient dissolved oxygen in water leads to the onset of anaerobic conditions and subsequently to the release of odourous or noxious gases such as H_2S. So the measurement of the D.O. in a water body is of particular importance when dissolved oxygen is sufficiently present in a water body organic materials, and wastes are degraded effectively. Under such conditions, carbon becomes CO_2, sulphur becomes – SO_4^-, phosphorus is converted to – PO_4^- and nitrogen forms ammonia and nitrates. On the other hand, when the dissolved oxygen is limited or insufficient to facilitate microbial activity, methane is released from carbon, odourous amines result from nitrogen and the foul-smelling – H_2S gas from sulphur.

The amount of D.O. also influences the profile of aquatic life. For example following table gives an estimate of the effect of decreasing D.O. in a river on the optimal productivity of fish. (Relationship between the dissolved oxygen of a water body and productivity of fish).

D.O. (mg/L)	% production (optimum)
5.0	100
4.5	90
4.0	75
3.5	40
3.0	0.0

Principle:

1. The dissolved oxygen can be determined by Wrinkler's Idometric titration method.
2. The principle of this titration based on the fact that the equivalent amount of iodine liberated that will be corresponds with oxygen percentage in the water sample.
3. The manganese sulphate ($MnSO_4$) reacts with alkali to form a white precipitate of manganese hydroxide which in the presence of oxygen gets oxidised to a brown colour compound.

4. In the strong acid medium Mn ions are reduced by iodine ions which gets converted into free iodine equivalent to the original concentration of oxygen in the sample. The free O_2 can be titrated against hypo solution (sodium thiosulphate) using starch as an indicator.

Chemical Reactions:

1. The method is based on oxidation-reduction reaction. In that the KOH reacts with $MnSO_4$ to give rise precipitate of manganese hydroxide. It is observed that oxygen is present.

$$MnSO_4 + 2KOH \longrightarrow Mn(OH)_2 + K_2SO_4$$
$$\text{White ppt}$$

2. But in the presence of oxygen the highly alkaline solution of manganese hydroxide is oxidised to form brown coloured precipitate of manganese oxyhydrate.

$$2\,Mn(OH_2) + O_2 \longrightarrow 2MnO\,(OH)_2$$
$$\text{Brown ppt}$$

3. This reaction occurs in direct proportion to the amount of oxygen present. Thus, approximate amount of O_2 may be proportional to the intensity of brown colouration or ppt. Addition of conc. H_2SO_4 dissolves the precipitate, forming manganese sulphate.

$$MnO\,(OH)_2 + \text{conc. } H_2SO_4 \longrightarrow Mn(SO_4)_2 + 3H_2O$$

4. There is intermediate reaction takes place between intermediate compound and potassium iodide which has been previously added. Liberating free iodine giving typical yellow colour to the solution.

$$Mn(SO_4)_2 + 2KI \longrightarrow MnSO_4 + K_2SO_4 + I_2 \uparrow$$

5. The amount of free iodine which is liberated equivalent to the amount of oxygen present in water. The iodine can be titrated against sodium thiosulphate using starch as an indicator.

$$2\,Na_2S_2O_3 + I_2 \longrightarrow Na_2S_4O_6 + 2NaI$$
$$\text{(Colourless)}$$

(a) Apparatus and Glassware:

Burette 50 ml capacity.

Reagent bottles with stoppers 250 ml capacity.

2-conical flasks

Measuring jar

Graduated pipette.

(b) Material:

River water and pond water for finding out oxygen content in each case.

(c) Chemical reagents :

(i) **Sodium thiosulphate ($Na_2S_2O_3$, 0.025 N):** Dissolve 6.20 gms of $Na_2S_2O_3$ in boiled distilled water and make volume upto 250 ml. Add 2 or 3 pallets of NaOH as stabiliser in the solution. This is 0.1N stock solution. Dilute it to 4 times with boiled distilled water to prepare 0.025 N. $Na_2S_2O_3$ solution. Keep the solution in brown glass bottle.

(ii) **Alkaline potassium iodide solution:** (Wrinkler - B): Dissolve 100 g of KOH and 50 g of KI in 200 ml of boiled distilled water.

   (iii)**Manganeous sulphate solution:** (Wrinkler's A solution): Dissolve 100 gms of $MnSO_4$ · $4H_2O$ in 200 ml of boiled distilled water and filter it.

   (iv)**1% starch solution:** Dissolve 1 g of starch in 100 ml of warm (80-90°C) distilled water and add a few drops of formaldehyde solution.

   (v) **Sulphuric acid:** H_2SO_4 conc. (specific gravity 1.84).

(d) Procedure:

1. Fill the sample in a glass stoppered bottle of known volume (100-300 ml) carefully, care should be taken to avoid any kind of bubbling and trapping of the air bubbles after placing the stopper.
2. And 1 ml of each $MnSO_4$ and alkaline KI solutions well below the surface from the walls of bottle. Use always separate pipettes for these two reagents. A brown precipitate will appear indicating the presence of oxygen.
3. Place the stopper and shake the content well by inverting the bottle repeatedly and keep the bottle for sometime to settle down the precipitate.
4. Add 1-2 ml of conc. H_2SO_4 and shake well to dissolve the precipitate.
5. Remove the 100 ml of this amount in a conical flask for titration.
6. Titrate the contents, with sodium thiosulphate (0.025 N) solution using starch as an indicator; at the end point, initial dark blue colour changes to colourless.
7. Repeat the same procedure for another sample.
8. Record the readings in tabulated form.

Observation Table:

(I)

Sr. No.	Burette reading in ml for river water	Mean
1.		
2.		
3.		

(II)

Sr. No.	Burette reading in ml for pond water	Mean
1.		
2.		
3.		

(e) Calculations:

The amount of dissolved oxygen can be calculated using following formula.

$$\text{Dissolved oxygen, mg/L} = \frac{x \times N \times 8 \times 1000}{V_2 \left(\dfrac{V_2 - V}{V_1} \right)}$$

where,

V_1 = Volume of small water after placing stopper (300 ml)
V_2 = Volume of part of content titrated (100 ml)
V = Volume of $MnSO_4$ and KI added - 4ml
X = Mean of Burette reading
N = Normality of sodium thiosulphate solution (0.025 N)

****

Practical **2**...

Estimation of Water Alkalinity from given Water Sample

Aim : To estimate alkalinity from given water sample.

Object: To estimate the given sample of water,

Principle:

Alkalinity of water sample refers to its capacity to neutralize acids. it is the sum total capacity of chemical components in water that tend to raise the pH of water above 4.5. Anionic radicals such carbonates, hydroxides and phosphates contribute to increase in alkalinity. It is expressed as mg/l of $CaCO_3$ of water. Alkalinity, is important for aquatic life because it acts as a buffer, controlling pH fluctuations.

Requirements: Conical flasks, phenolphthalein indicator, N/50 H_2SO^4, methyl orange indicator and burette.

Procedure: For estimating alkalinity of the given water sample proceed as under:

1. **Phenolphthalein alkalinity:** Take 100 ml of given water sample in a conical flask add 4 drops of phenolphthalein indicator in it. If colour appears in solution, it contains hydroxide normal carbonate. If sample becomes pink then add N/50 sulphuric acid through a burette until pink colour just disappears. Note the ml of acid used and calculate as under:

Calculation: The phenolphthalein alkalinity, expressed in ppm of $CaCO_3$, is equal to:

$$\text{ml of N/50 } H_2SO_4 \text{ used} \times 10$$

2. **Methyl orange alkalinity:** Take 100 ml of given water sample in a conical flask and add 2 drops of methyl orange indicator in it. If yellow colour develops in solution, it contains carbonate or bicarbonate. Then add N/50 sulphuric acid through a burette until yellow colour shows a first change from pure yellow to faint orange tint. Stop dropping H_2SO_4 and note the ml of acid used.

Calculation: The methyl orange alkalinity, expressed in ppm of $CaCO_3$, is equal to

$$\text{mL of N/50 } H_2SO_4 \text{ used} \times 10$$

***

Practical **3**...

Study of Animal Community Structure by Quadrat, Line and Belt Method

Aim: To study of Animal Community Structure by Quadrat, Line and Belt Method.

Introduction:

Plants growing together have, mutual relationships among themselves and with the environment. Such a group of plants in one area forms a stand. Several similar stands represent a community which is a part of an ecological system (the ecosystem) in which transformation, accumulation, and flow of energy are involved. The functioning of this system is intimately related with the components of the community. The components vary in quality as well as in quantity and impart a structure to the community.

Community Structure:

The structure of a community can be studied by taking into consideration a number of characters which are usually grouped under two heads viz. analytic and synthetic. Certain analytic characters viz. frequency, density, abundance and dominance can be expressed quantitatively while others viz. sociability, vitality, periodicity and stratification find only qualitative expression. Synthetic characters include presence, Constance and fidelity of components and may be computed from analytic characters of several stands of a community.

The analytic characters of a community are determined by means of three main sampling units—area, line and point, as employed in quadrat, transect and point method respectively.

Objectives:

After doing this experiment you will be able to:

- describe about quadrat, line and belt methods and perform them.
- collect and identify the living organisms of the particular population.
- point out which is the most effective sampling method and how to use it properly in any given area.
- know the limitation of apparatus and techniques you are going to use.

Material required:

Quadrats, pegs, thread, record book, magnifying glass, gloves.

Methods:

Quadrats: These have been used extensively in determining the distribution of plant communities but can also be used with slow moving invertebrates such as those which occur in leaf litter or in intertidal habitats.

Quadrats arc sampling units of a known area. Now you must be thinking why sampling is necessary? It is seldom possible to count all the individual animals or plants within a given population. This would not only be extremely laborious and time consuming. But would almost certainly involve disturbance and damage to the habitat and population we wish to study. Thus by sampling, we aim to select for study a small representation of the total population. These sample units must be distinct, must not overlap and together they make up the total population. The number of individual of a species in each sampling unit is then counted or estimated and from this information you can obtain frequency and distribution of that species in the population as a whole.

Now, let us study about the Structure of quadrats. Usually the quadrats have a rectangular frame (Fig. 2.1) and come in a variety of dimensions. If frame is not available then you can make a quadrat of known dimension by using 4 pegs and string or strong thread. The pegs are inserted in the ground at four comers with equal distance from each other (Fig. 2.2). The distance dimension may be decided by oneself.

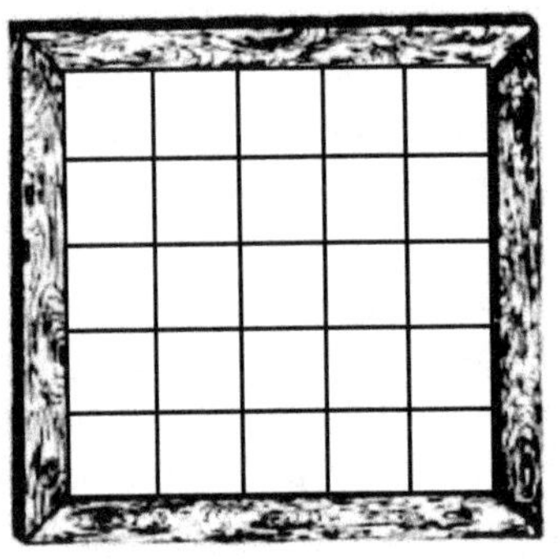

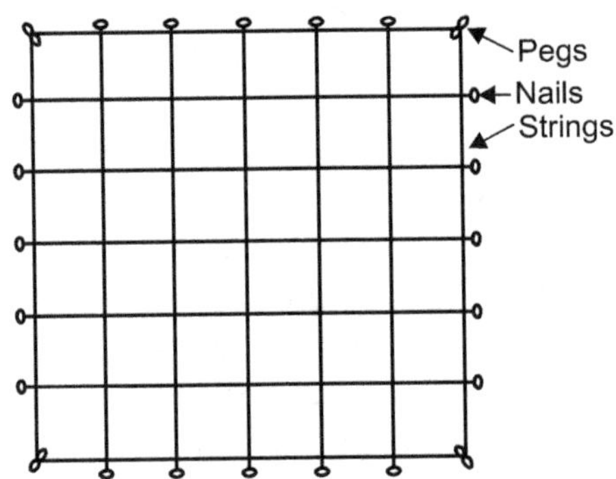

Fig. 2.1: Quadrat frame. 0.5 cm wooden frame with wires fixed at 10 cm intervals

Fig. 2.2: Quadrat made from 4 pegs, strings and nail, fixed at equal distance and squares are made with help of a string

When you use quadrat it is assumed that its contents will represent the whole sampling area. More commonly used are 1 m^2 and 0.25 m^2 frame quadrats. These are easily constructed out of wood and with cross wires or strings subdividing them at 10 cm intervals for the ease of counting.

Now, the question arises, how the size of the quadrat will be determined. If the dispersion of a population within the sampling area is truly random, then all the quadrat sizes would be equally efficient in the estimation of that population. However, the spatial dispersal of a population is seldom random or regular. An aggregated distribution (Fig. 2.3) is more likely with individuals found in paths. This is because several environmental factors will be unevenly distributed within a sampling area.

Fig. 2.3: An aggregate distribution

Generally small size quadrat has been found to be more efficient than a large one when population is aggregated. Instantly the question comes to mind is why? The reasons are following:

*   More small samples can be taken for the same amount of labour.
*   More number of small quadrats cover a wider range of habitat than a few large ones and the sample will be, more representative.
*   Statistical error will be reduced as a sample of many small units will have more degree of freedom than a sample of a few large units.

Although a small quadrat may be theoretically best, there are practical considerations which set a lower limit on the size. Thus when sampling in a wood a small quadrat may undersample the dominant species of tree. In addition, the smaller the quadrat the greater the sampling error at its edges; are the plants on the edges of the quadrat to be included or not?

Thus, for determining the optimum quadrat size for a particular type of vegetation, a series of quadrats of increasing size are laid out. The cumulative number of plant species counted after each successive increase in quadrat size is then recorded e.g.

Quadrat size	Total number of species
0.025	10
1	14
4	19
8	22
16	25

Eventually a point is reached where a further large increase in-quadrat size results in only a few extra species. Since the common species will have already been included, the extra time and effort required in recording very large quadrat is unproductive.

The optimum quadrat size is reached when a 1% increase in quadrat size produces no more than a 0.5% increase in the number of species present (Fig. 2.4).

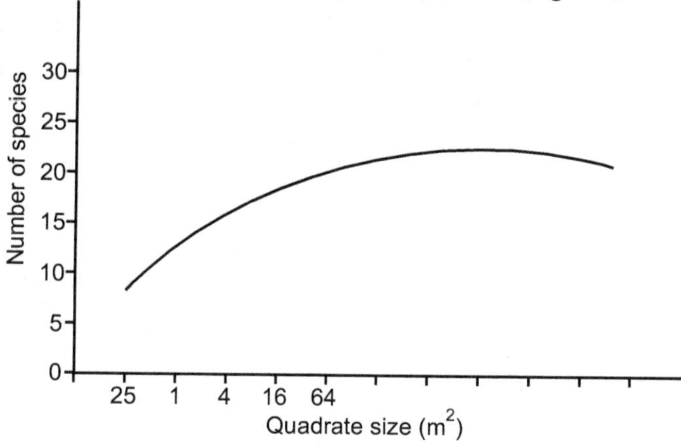

Fig, 2.4: Graph to determine the optimum quadrat size

Then comes the number of quadrats as you have seen around yourself that a large variation is found when sampling in natural populations. In order to make our result statistically significant a large number of samples should be taken. However, sorting and counting all the species in a very large sample can be tedious and time consuming. A similar exercise enables us to estimate the optimum number of quadrats, required when studying the species composition of a particular site. A series of quadrats of satisfactory minimum - size is placed randomly across the sampling area. The cumulative number of species is recorded after each increase in quadrat number used; for example,

Number of Quadrats	Total number of species
1	14
4	34
8	37
16	40

Eventually a point is reached when all the common species have been identified and a further increase in quadrat number will not merit the time and effort required (Fig. 2.5). A satisfactory minimum number of quadrats is reached when a 1% increase in the number of quadrats has no more than a 0.5% increase in number of species found.

When the size and number of quadrats is known, we can study the structure of community by simply making the necessary observations. By the above described method you can make a quadrat of definite size by string and pegs or wooden quadrat of known size. The number of quadrats is already known.

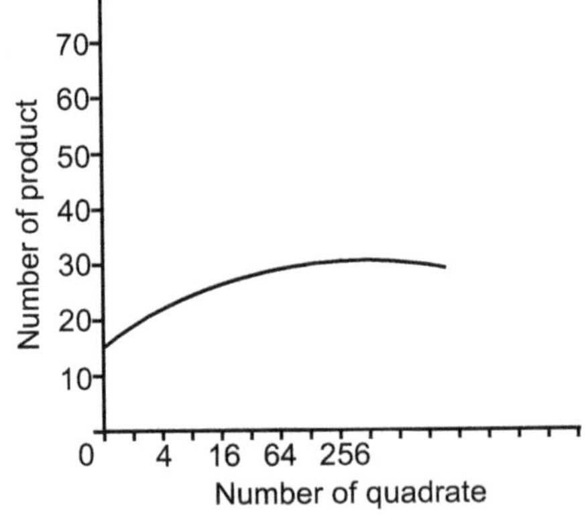

Fig. 2.5: Graph to determine number of quadrats

Now, you can make observations and record them in your notebook in the form of a table given below.

Sr. No.	Name of the Species	Quadrat laid down										
		1	2	3	4	5	6	7	8	9	10	11
1	A	–	–	–	–	–	–	–	6	–	4	–
2	B	4	2	1	–	–	–	–	3	–	2	–
3	C	1	8	–	10	–	–	–	11	7	8	2

Transects:

The use of transects constitutes a form of systematic sampling but in this case the samples are studied in a heat fashion. Transects are useful for recording changes in the species composition of plant communities where some sort of transition exists, e.g. from water to land or from one soil type to another. Commonly two types of transects are used:

1. Belt transect:

This is a strip usually 0.5 m in width that is located across the study area in such a way as to highlight any transition. A tape or rope marked off at 0.5 m intervals is laid across the sampling area and a 0.25 m^2 frame quadrat laid down at 0.5 m intervals by the side of the tape or rope to give a continuous belt transect. The animals and plants within each quadrat are identical and counted and an estimation made of their relative abundance. This procedure with a transect over 15 metres long is time consuming and it is more usually to carry out quadrat sampling at every metre interval to give a ladder transect.

As by doing this experiment you can see that this sort of sampling is quite intensive but it gives an accurate record of the organisms present. However, the length of the transect will be limited by the time available. The transect work will also require the recording of a profile indicating changes in the height of the ground.

Data Sheet for Belt Transect

1	2	3	4	5	6	7	8	9	10	11	12	13	14	15

Name of species	Quadrat number
....	....
....	....
....	....
....	....
....	....
....	....
....	....
....	....
....	....
....	....

Line Transect Method:

This method takes less time and is less quantitative and therefore, less representative. A tape or rope marked off at 0.5 m intervals is laid along the area of ground to be sampled. The plant species that are touched or covered by the line are then recorded either all the way doing or else at regular intervals. There may be many species which do not touch the line and are therefore not recorded, thus the results may give a completely unrealistic sample of the community on which the line transect is made. However, if you want an impression of the main features of a transition in all, then this method has considerable use. The data sheet will be the same as belt transect method.

Observations and Results:

Observations should be entered in the given data sheet by recording + or – or the number species are recorded. The species which is present in several quadrats is abundant and species present in few quadrats is rare one.

Precautions:

The most important precaution to keep in mind is conservation of the environment. The following points should be kept in mind:

1. The number of individual animals and plants collected should be kept to a minimum, depending upon the nature of the experiment. If there is need for quantitative study then number of sample taken should be strictly limited.

2. As far as possible all organisms removed during the exercise should be returned to their original habitat after the experiment is over.

3. The collection should be avoided from the same site as it would have adverse effect upon the density of plants and animal species within the community in question.

4. During collection, damage to the habitat should be avoided because it causes adverse effect on organisms found there.

Discussion:

Now it should be clear that in order to determine the structure of community you have quardrat, belt transact and line transact methods. Now, you have to decide which of the methods would be followed to study the community structure.

The next, and major problem is the problem of identification of plants and animals. The identification of plant and animals down to species level can be difficult and very time consuming. It will be good if you can identify the species by using prepared illustrations and diagnosis key. But this is difficult task as you have limited time. But the counsellor at your study centre will help you in identifying the plants and animal species.

***

Practical **4**...

Determination of Density, Frequency and Abundance of Species by Quadrat Method

Aim: To determination of density, frequency and abundance of species by quadrat method.

Object: To study the community by quadrat method by determining frequency, density and abundance of different species present in the community.

Requirements: Quadrat of definite size (50 cm × 50 cm in the present case) and meter scale.

Procedure: For determining the frequency, density and abundance of different species in a community proceed as under:

Throw a quadrat of lay down a quadrat in a specified area of the grassland, forest or garden or pond or any place where community is to be studied. Note carefully each type of animals or plants and or both, as needed, occurring in the squares of the quadrat. Note down the number and names of individual types of animals and plants. Repeat it by throwing or laying down the quadrat randomly in the area atleast 10 times. Record your data in form of the following table.

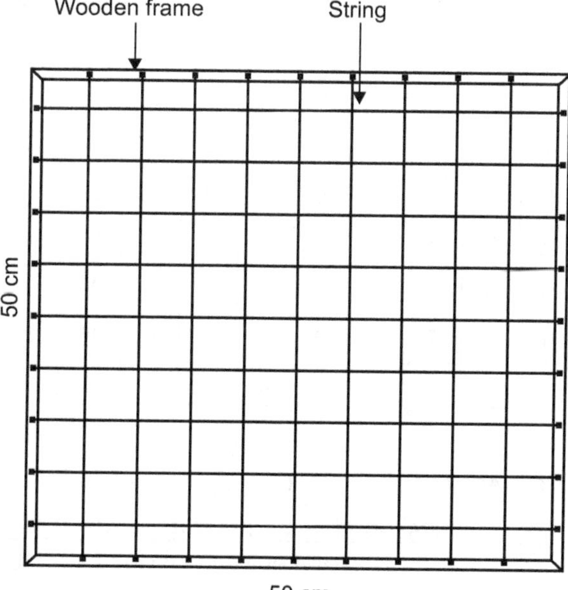

Fig. 2.6: A Wooden Quadrat 50 cm × 50 cm

Sr. No	Name of species	Number of individuals in each quadrat										Total number of individuals of species	Total number of quadrats occurrence	Total number of quadrats studied	Frequency %	Density (Per unit area)	Abundance
		1	2	3	4	5	6	7	8	9	10						
1	2	3										4	5	6	7	8	9
1.	A	0	4	0	0	6	0	0	0	0	0	10	2	10	20	4.0	5
2.	B	0	0	0	5	3	0	0	0	0	3	11	3	10	30	4.4	3.6
3.	C	4	0	0	0	0	0	6	0	0	0	10	2	10	20	4.0	5
4.	D	1	0	0	1	0	1	0	0	1	0	4	4	10	40	1.6	1
5.	E	2	2	2	1	3	3	1	1	1	2	18	10	10	100	7.2	1.8
6.	F	10	7	15	11	20	14	5	18	11	6	117	10	10	100	46.8	11.7
7.	G	5	0	0	4	0	0	0	6	0	0	15	3	10	30	6.0	5
8.	H	6	4	5	10	8	7	5	10	10	2	67	10	10	100	26.8	6.7
9.	I	2	3	4	2	6	0	3	2	8	2	32	9	10	90	12.8	3.5
10.	J	1	0	5	7	3	6	4	7	3	2	38	9	10	90	15.2	4.2
11.	K	6	4	0	5	3	0	2	1	7	3	31	8	10	80	12.4	3.8
12.	L	8	10	12	0	0	8	6	0	4	1	49	7	10	70	19.6	7
13.	M	0	6	4	3	2	1	0	0	0	0	16	5	10	50	6.4	3.2
14.	N	3	0	0	0	0	0	0	0	0	0	3	1	10	10	1.2	3

Calculation:

A. Frequency: The % frequency is determined by the following formula:

$$\text{Frequency \%} = \frac{\text{Total number of quadrats in which species occurred}}{\text{Total number of quadrats studied}} \times 100$$

$$\text{Or} \quad \frac{\text{Value of column 5}}{\text{Value of column 6}} \times 100$$

The frequency is the number of sampling unit or quadrats in which a particular species occurred.

B. **Density:** the density is determined by the following formula:

$$\text{Density} = \frac{\text{Total number of individuals of species}}{\text{Total number of quadrats studied}}$$

$$\text{Or} \quad \frac{\text{Value of column 4}}{\text{Value of column 6}}$$

The density of a particular species is the number of its individuals occurring per unit area.

The density values, thus, obtained from the above formula for each species, are to be expressed as individuals per unit area. For example, calculated value for species A according to above formula is 1. This value is in 2500 cm^2 area i.e. the area of the quadrat used 50 $\times$ 50 cm^2. Therefore, in 2500 cm^2 area there is 1 individual.

Hence, 1 cm^2 area there would be $\dfrac{1}{2500}$ individuals.

And in 100 $\times$ 100 cm^2 i.e. 10000 cm^2 or (1 m $\times$ 1 m, i.e. 1 m^2).

$$= \frac{10000 \times 1}{2500} = 4$$

Thus, density of species A is to be expressed as 4 per m^2 instead of 1.

C. **Abundance:** The abundance is determined by the following formula:

$$\text{Abundance} = \frac{\text{Total number of individuals of species}}{\text{Total number of quadrats of occurrence}}$$

$$\text{Or} \quad \frac{\text{Value of column 4}}{\text{Value of column 5}}$$

The abundance may be described as number of individuals occurring per quadrat.

***

Practical **5**...

Study of Microscopic Fauna of Fresh Water Ecosystem

Aim : To Study of microscopic fauna of freshwater pond.

Requirements:

Compound microscope, slide, coverslip, glacial acetic acid, dropper, plastic cans, stopper bottles, blotting paper, centrifuge machine, hand net.

Procedure:

1. Take two clean, dry plastic cans of 5 litres capacity.

2. Collect the water sample in one can from polluted water and one from non-polluted water.

3. Collect the macro organisms with the help of hand net from both the localities and bring the samples and organisms in the laboratory.

4. Take the drop of water with glass dropper on a clean and dry slide and place the coverslip on it. Remove excess of water with the help of blotting paper and observe under low power. Few moving micro-organisms will be seen. For getting more organisms centrifuge larger quantity of sample and remove the supernatant. Take the drop from sediment. Put coverslip you will find large number of organisms.

5. You will find floura and fauna. Make the list of organisms according to their phyla.

6. Observe their body movements, and sketch the figures of the animals. Refer the figures from text book for the purpose of identification of animal.

7. If you want to see the detailed structure of animal use glacial acetic acid drop for ceasation of the movements. Animal may get fixed. For this purpose use very dilute acetic acid. Use high power whenever necessary.

8. Repeat the procedure 2-3 times and try to record different types of forms.

9. Make the list of organisms found in polluted and unpolluted water samples and compare.

Table 2.1: Fauna from polluted and unpolluted water samples

Sr. No.	Name of the animal from polluted sample	Name of the animal from unpolluted sample
1.	Amoeba	Euglena
2.	Pelomyxa	Pelomyxa
3.	Actinophrysts	–
4.	Colpoda	Colpoda
5.	Paramecium	Paramecium
6.	Vorticella	Daphnia
7.	Euplotes	Hydra
8.	Oxytrica	Lymnea
9.	Rotifers	Mosquito larvae
10.	Nematod worm	Shriump
11.	Daphnia	Bivalve
12.	Ostracod	Notonecta
13.	Cyclops	Water strider
14.	Ceratula	Dragonfly nymph
15.	Sylonychia	Damselfly nymph
16.	Gastrostyla	Cyclops
17.	Planorbis	Cypris
18.	Lymnea	Corixa
19.	Ranatra	Nepa
20.	Mosquito larvae	
21.	Gambusia	

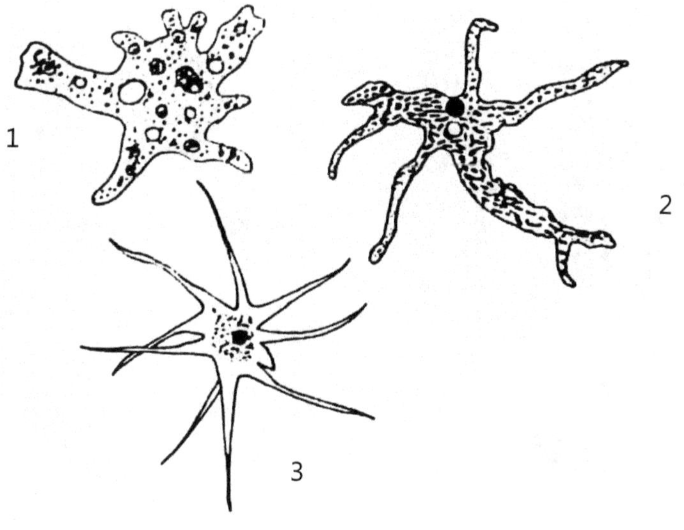

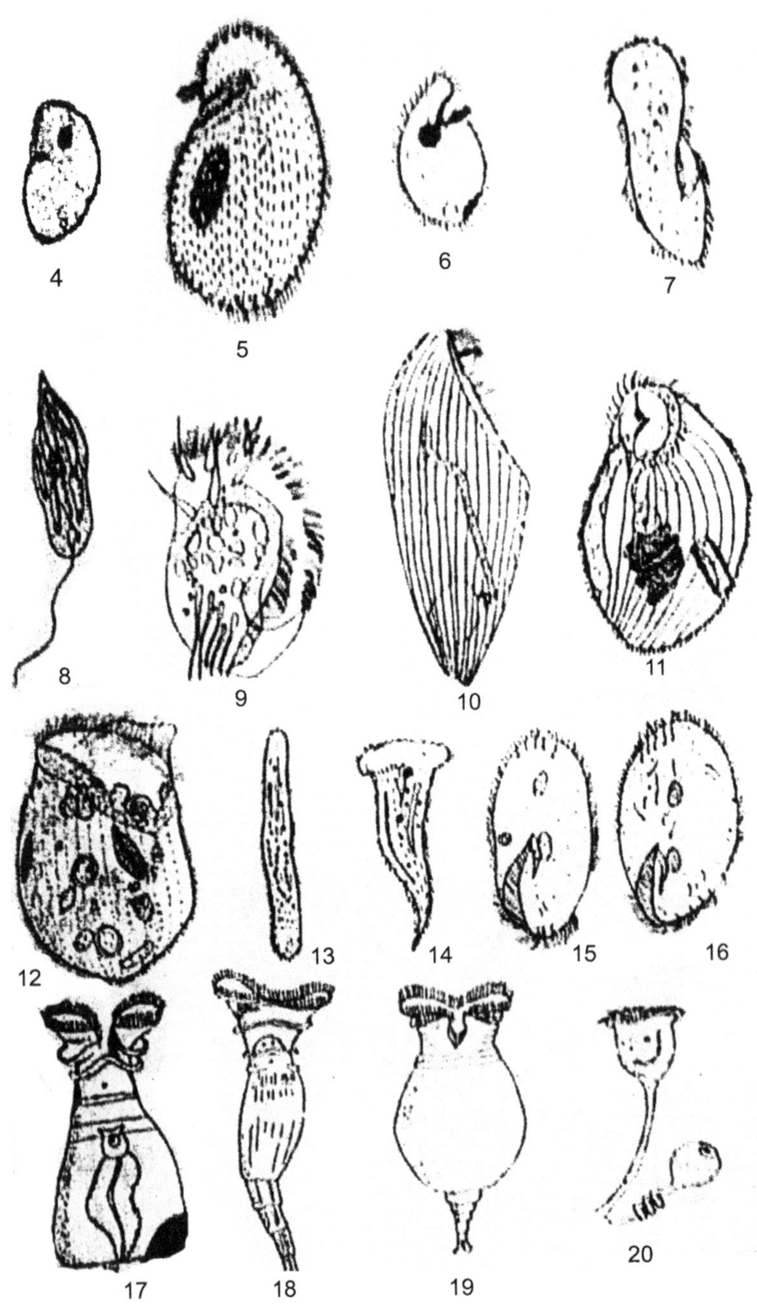

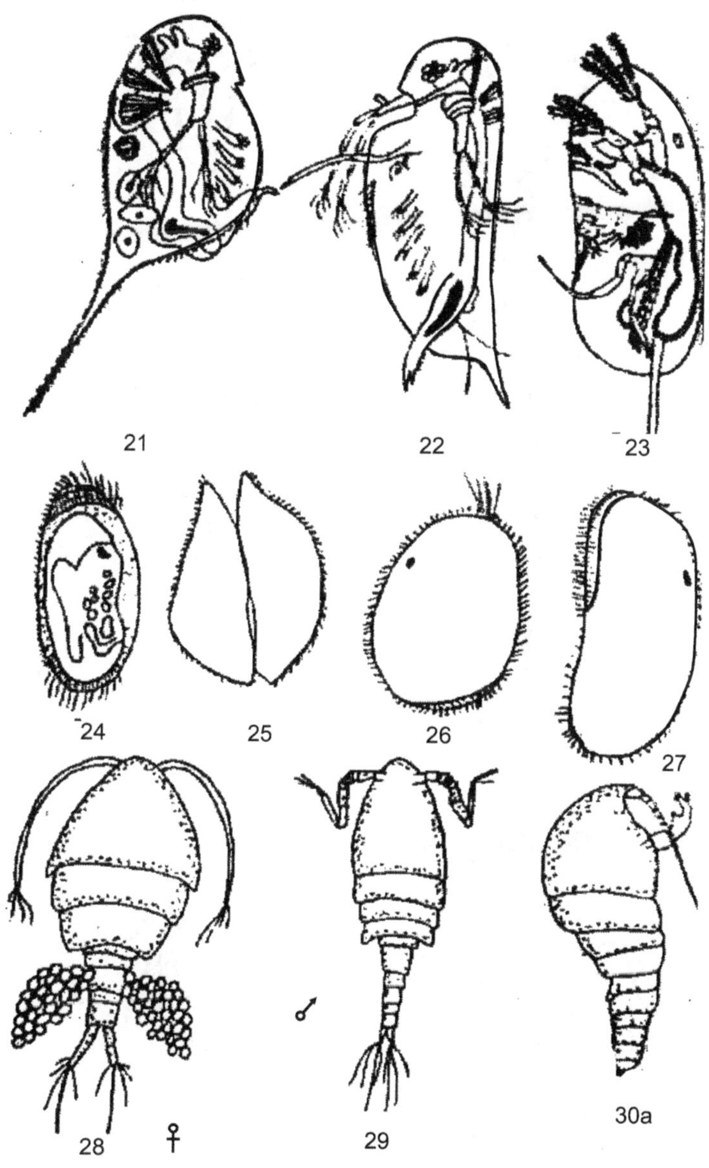

21
22
23
24
25
26
27
28
29
30a

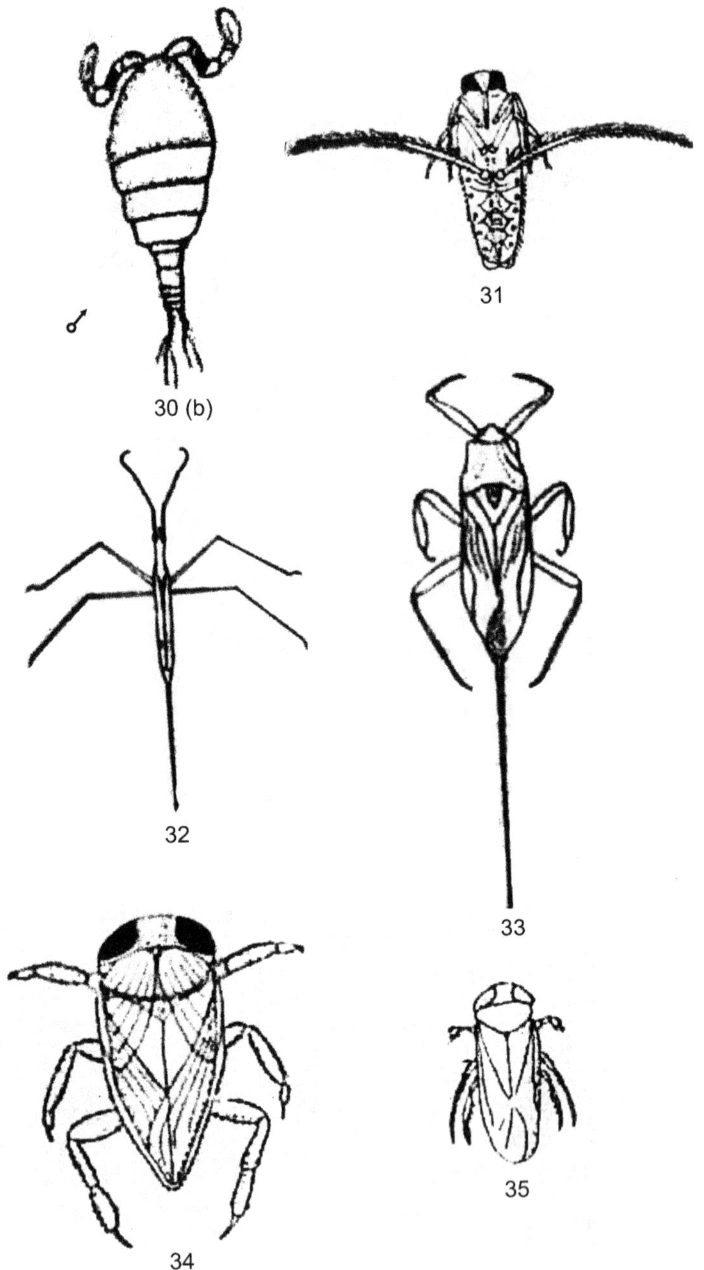

Fig. 2.7; Fauna

1. *Amoeba proteus = now chaos diffluens)*
2. *Amoeba discoides*
3. *Amoeba radiosa (= Astramobae)*
4. *Colpoda inflate*
5. *Plagiophly nasuta*
6. *Colpoda aspera*
7. *Paramoecium sp*
8. *Euglena sp*

<table>
<tr><td>9.</td><td>*Euplotes charon*</td><td>10.</td><td>*Blepharisma intermedium*</td></tr>
<tr><td>11.</td><td>*Condylostoma*</td><td>12.</td><td>*Condylostoma certicella*</td></tr>
<tr><td>13.</td><td>*Spirostomum ambiguum*</td><td>14.</td><td>*Stentor sp*</td></tr>
<tr><td>15.</td><td>*Oxytricha oblongatus*</td><td>16.</td><td>*Oxytricha ovalis*</td></tr>
<tr><td>17.</td><td>*Philodina citrina (Rotifer)*</td><td></td><td></td></tr>
<tr><td>18.</td><td>*Philodina sp. with expanded trochal disc*</td><td></td><td></td></tr>
<tr><td>19.</td><td>*Rotaria valgaris*</td><td>20.</td><td>*Verticella sp*</td></tr>
<tr><td>21.</td><td>*Daphnia carinata (Female)*</td><td>22.</td><td>*Daphnia carinata (Male)*</td></tr>
<tr><td>23.</td><td>*Ostracod (Female)*</td><td>24.</td><td>*Stenocypris malacomsomi*</td></tr>
<tr><td>25.</td><td>*Cypris shell*</td><td>26.</td><td>*Centocypris shell*</td></tr>
<tr><td>27.</td><td>*Heterocypris shell*</td><td>28.</td><td>*Mesocyclops hyalinus (Female)*</td></tr>
<tr><td>29.</td><td>*Mesocyclops hyalinus (Male)*</td><td></td><td></td></tr>
<tr><td>30.</td><td>*Mesocyclops leuckart (Female and Male)*</td><td></td><td></td></tr>
<tr><td>31.</td><td>*Notonecta glauca*</td><td></td><td></td></tr>
<tr><td>32.</td><td>*Ranatra elongate*</td><td></td><td></td></tr>
<tr><td>33.</td><td>*Laccotrephes maculates*</td><td></td><td></td></tr>
<tr><td>34.</td><td>*Corixa sp.*</td><td></td><td></td></tr>
<tr><td>35.</td><td>*Sigara pectoralis.*</td><td></td><td></td></tr>
</table>

✳✳✳

Practical **6**...

Estimation of Water Holding Capacity of given Solid

Aim : To Estimate the water holding capacity of given soil.

Principle :

The water holding capacity of a soil is very important agronomic characteristic. Soils that hold generous amounts of water are less subject to leaching losses of nutrients or soil applied pesticides. This is true because a soil with limited water holding capacity (i.e. a sandy loam) reaches the saturation point much sooner than the soil with a higher water holding capacity (i.e. a clay loam). After a soil is saturated with water, all of the excess water and some of the nutrients and pesticides that are in the soil solution are leached downward in the soil profile. Soil texture and soil organic matter controls the water holding capacity of soil. In general, the higher percentage of silt and clay size particles, the higher the water holding capacity. The small particles (clay and silt) have a much larger surface area than the larger sand particles. This large surface area allows the soil to hold a greater quantity of water.

Requirements: Two soil samples, two measuring cylinders, two funnels, filter paper.

Procedure:

1. Take 2 funnels and line them with filter paper.
2. Put these funnels in a measuring cylinder.
3. Take the two soil samples like roadside soil and garden soil. The weight of soil sample should be same in both (50 gm).
4. Put this soil into two funnels separately.
5. Take 50 ml of water and pour equal amount of water in both funnels.
6. Let the water drip in the cylinder.
7. After 30 minutes note the volume of water collected.

Observations:

The volume of water collected in cylinder of sample A was more than the sample B.

Calculations:

Volume of water retained by soil = Volume of H_2O poured − Volume of H_2O collected in cylinder.

Water holding capacity of soil =

$$\frac{(\text{Volume of } H_2O \text{ retained})}{\text{Volume of } H_2O \text{ required}} \times 100$$

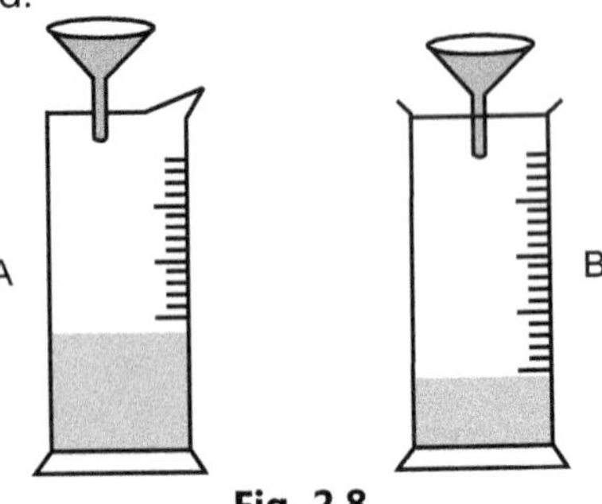

Fig. 2.8

Conclusion:

Soil sample A is road side and sample B is garden soil as sample B has more water holding capacity.

Precautions:

The water should be poured slowly. Measuring cylinders should be properly calibrated.

***

Practical **7**...

Estimation of Dissolved and Free Carbon Dioxide from Water Sample

Aim : To Estimate the dissolved and free carbon dioxide from water sample.

Introduction :

Surface water contain normally less than 10 mg/lit of free CO_2 while some ground wall exceed this concentration. Though carbon dioxide is readily soluble in water. Very little CO_2 in a water sample is because of the small amount of it being present in atmospheric air. A part of the decomposition of the small organic matter and the respiration of aquatic plants and animals contribute to free CO_2 present. CO_2 combine chemically with water to form carbonic acid (H_2CO_3) which affects pH of water. Much of CO_2 is present in the form of bicarbonate and carbonates.

Principle :

Free CO_2 reacts with strong alkali i.e. NaOH or Na_2CO_3 (usually carbonate free) to form bicarbonates ($NaHCO_3$) at pH 8.3. Completion of this reaction is indicated potentiometrically characteristics of phenolphthein indicator.

Apparatus :

25 ml burette, 5 ml pipette, 250 ml conical flask and etc.

Polluted and unpolluted pond water sample.

Chemical Reagents :

1. **Sodium hydroxide, (0.05 N):** Prepare 1.0 N NaOH solution by dissolving 40 g of NaOH in CO_2 free distilled water (boiled) and make the total volume 1 litre. Dilute 50 ml of 1N NaOH to 1 litre; so as to become 0.05 N NaOH (standardise it with H_2SO_4 or oxalic acid).

2. **Phenolphthalein Indicator:** Dissolve 0.5 g of phenolphthalein in 50 ml of 95% ethanol and add 50 ml of distilled water.

Procedure:

1. Fill the burette with 0.05 N NaOH.

2. Take 100 ml of unpolluted water sample in a conical flask.

3. Add few drops of phenolphthalein indicator. If the colour turns pink, free CO_2 is absent.

4. If the sample remains colourless titrate it against 0.05 N NaOH.

5. At the end point a pink colour appears. Record the readings (3) in tabulated form.

6. Repeat the same procedure for polluted sample of water.

Observation Table :

Type of sample		Burette readings in ml	Mean reading in ml
(i) Unpolluted	(a)		
	(b)		
	(c)		
(ii) Polluted	(a)		
	(b)		
	(c)		

Calculation :

$$\text{Free } CO_2,\ \text{mg/L} = \frac{(ml \times N) \text{ of NaOH} \times 1000 \times 44}{ml \text{ of sample}}$$

where,   N = Normality of NaOH

        41 = Molecular weight of CO_2

        1000 = per litre (for 1000 ml of water)

****

Practical **8**...

Study of Eutrophication in Lake/River

Aim : To study Eutrophication in lake/river.

Eutrophication (from Greek *eutrophos*-well nourished) or hypertophication is when a water body becomes overly enriched with minerals and nutrients which induce excessive growth of algae. This process may result in oxygen depletion of the water body. One example is an 'algal bloom' or great increase of phytoplankton in a water body as a response to increased levels of nutrients.

Eutrophication is often induced by the discharge of nitrate or phosphate containing detergents, fertilizers or sewage into an aquatic system.

Eutorphication most commonly arises from the oversupply of nutrients, most commonly as nitrogen or phosphorus, which leads to overgrowth of plants and algae in aquatic ecosystems. After such organisms die, bacterial degradation of their biomass results in oxygen consumption, there by creating the state of hypoxia. The larger aquatic animals like fishes are killed due to suffocation.

- The sources of excess of phosphates are phosphates in detergent, industrial/domestic run-offs, and fertilizers.
- Excess nutrients in the form of fertilizers are applied to soil and they get drained into water body.
- The excess nutrient cause an algal boom. The algal bloom blocks the sunlight reaching the bottom of water body.
- Plants beneath the algal bloom die because of lack of sunlight for photosynthesis.
- Eventually the algal bloom dies and sinks to the bottom of the lake.
- Bacteria begins to decompose the remains using up oxygen for respiration.
- The decomposition causes the water to become depleted of oxygen.
- Larger life forms, such as fish, suffocate to death.
- This water body can no longer support life.
- When natural eutrophication is speed up by human activity, it is called cultural eutrophication. By clearing law and builidings of towns and cities, land runoff is accelerated and more nutrients like Phosphates and nitrates are supplied to lakes and rivers.
- Extra nutrients are also supplied by treatment plants, golf courses, fertilizers, farms (including fish farms) as well as untreated sewage in many countries.
- Enhanced growth of aquatic vegetation or phytoplankton and algal blooms disrupts normal function of ecosystem.
- Lack of oxygen needed for fish and shell fish to survive.
- Water becomes cloudy, coloured, green, yellow, brown or red.

- Value of rivers/lakes decreases.
- Health problems can occur where Eutrophic conditions interfere with water treatment.
- Some algal blooms are harmful and are toxic to plants and animals.
- If toxic fish, shellfish cause human food poisoning.

Prevention and Reversal:

1. Removal of phosphorous from rivers and lakes.
2. To restore shell fish populations in estuaries Oyster reefs remove nitrogen from water column and filter out suspended solids.
3. Seaweed farming useful for absorbing Phosphorus and nitrogen.
4. Minimizing non-point pollution.
5. Organic farming reduces harmful nitrogen leaching.
6. Dredging helps in reduction of soil contamination that was caused by sewage sludge, wilted plants or chemical spill.

For study of Eutrophication of river or lake arrange visit to such location and observe the following thing :

1. Take the sample of water containing algal bloom.
2. Identify the type of algae and sketch the figures.
3. Examine the aquatic microscopic fauna from the collected sample.
4. Take the water sample for estimation of dissolved oxygen and carbon dioxide.
5. Measure the pH of the water.
6. Measure the alkalinity, turbidity of the water.
7. Observe and record the colour and temperature of the water.
8. Note the sources of pollutants reaching to the river or lake.
9. If any fish kill in the lake or river record it.
10. Check whether the water from river or lake is used by the people living nearby.
11. Prepare the report after observations.
12. Make your suggestions to reduce the eutrophication of the lake or river.

✹✹✹

SECTION - III : ANIMAL DIVERSITY - II
(Semester - II)

Practical **1**...

Museum Study of Phylum Aschelminthes
Ascaris lumbricoides

Aim : Museum Study of Phylum Aschelminthes.

(a) PHYLUM – *ASCHELMINTHES*

Name of phylum derived from two Greek words; (askos = cavity or bag and helmins = worm) means the cavity between the digestive tube and the body wall.

General Characters :

1. Unsegmented body, triploblastic.
2. Body elongated, cylindrical or flattened and bilaterally symmetrical.
3. Body covered with thick, flexible cuticle with spines and bristles.
4. Epidermis syncytial or cellular.
5. Size small microscopic, a few meter or more in length.
6. Mostly aquatic, freshwater or marine, while some terrestrial free.
7. Anterior end carrying the mouth but head is not differentiated.
8. The respiratory and circulatory systems are absent.
9. Between the body wall and the digestive tract is a cavity which is not lined by mesoderm called *pseudocoel* instead of coelom.
10. Excretory system including canals and *protonephridia*.
11. Nervous system is simple, comprises anterior brain or circumenteric ring with anterior and posterior longitudinal nerves.
12. Sexes are separate (dioecious). Male usually smaller than the females.
13. Life cycle may be simple or complex.

Phylum Aschelminthes has one class – Nematoda.

1. ASCARIS LUMBRIOCOIDES

Systematic Position :

Phylum	-	**Aschelminthes**	-	Triploblastic; organ grade organisation; bilaterally symmetrical; pseudocoelomate
Class	-	**Nematoda**	-	Unsegmented body; cilia is absent, alimentary canal straight
Order	-	**Ascaroidea**	-	Buccal capsule absent mouth with three lips.
Genus	-	*Ascaris*		
Species	-	*lumbriocoides*		

Habit and Habitat :

It occurs in the intestine of man and pig.

Distribution :

Ascaris has world wide distribution.

Comments :

1. It is commonly known as *round worm*.
2. It shows sexual dimorphism. The males are smaller than females. The male measures 15-30 cm in length with a posterior curved tail. The female measures 20-35 cm in length and the posterior end is straight and blunt.
3. The body is elongated, cylindrical and some what pointed at both the ends.
4. Body surface is marked with four lines, these are mid-dorsal, mid-ventral and two lateral lines.
5. Mouth is situated at the anterior end which is guarded by three lips.
6. Excretory pore is small and lies on the ventral side about 2 mm away from anterior end.
7. Male is provided with curved tail with a pair of curved spicules known as *penial setae*.
8. Female genital aperture lies about 1/3rd of the length of body from anterior ends.
9. No intermediate host in the life history.

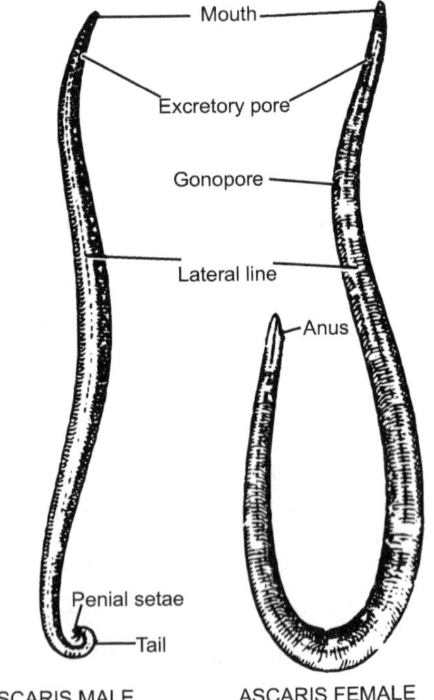

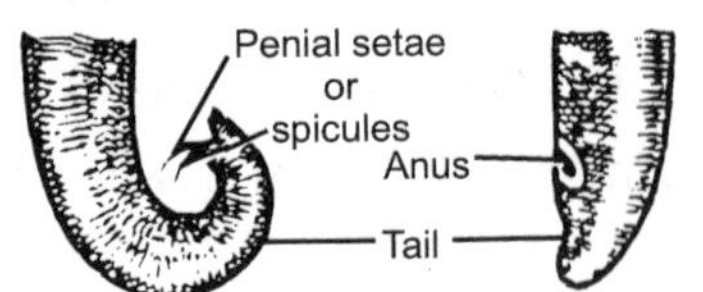

Fig. 3.1 : *Ascaris*

Ascaris is cosmpolitan in distribution. Specially prevalent in the tropics such as China, India and South-East Asia. The incidence of infection is highest in Korea and Philippines (i.e. 50-70%). It occurs in persons with unhygienic habits.

The adult worms is endoparasite in the small intestine of man. It is more common in children than adults. Their number may be 1000-5000 in single host.

The body of worms is elongated, cylindrical and tapering at both the ends. The body is covered by tough and elastic cuticle producing anti-enzymes. Colour of adult worms is pinkish white or yellowish white. Sexes separate.

Pathogenicity :

Infection is effected by swallowing ripe. *Ascaris* eggs with raw vegetables cultivated on a soil fertilized by infected human excreta. Infection may be through contaminated drinking water. The disease produced by adult Ascaris is known as *Ascariasis*. The symptoms after infection are of two types : (a) Those produced by migratory larvae and (b) Those produced by adult worms.

(a) Symptoms produced by larvae :

Due to heavy infection to lungs, typical symptoms of pneumonia such as fever, cough and dyspnoea may appear. The blood stained sputum contains *Ascaris larvae. Urticarial rash* and *eosinophilia* (20%) are seen in such cases. When larvae enter the general blood circulation, they produce disturbance to various organs due to their presence in the brain, spinal cord, heart and kidneys.

(b) Symptoms produced by Adult worms :

The adult worms live in intestine and produce pathogenic effects of gastro-intestinal tract. The worms take the digested food material from intestine causing deficiency of Vitamin - A (night blindness) and proteins, children suffer from malnutrition. Antienzymes (antitryptic and antipeptic) are liberated by *Ascaris* protect the worm from being digested by host's intestinal enzymes, they also help the development of malnutrition. The body fluid of *Ascaris* called *ascaron* or *ascarase* is toxic and irritating to the host. It give rise to typhoid-like fever, various allergic manifestations – such as urticaria, oedema of face, conjunctivitis and irritation of upper respiratory tract.

The adult worm also cause mechanical damage i.e. intestinal obstruction (in children) and penetration through the alimentary canal leading ulcers. When they reach in appendix cause appendicitis or in biliary passages cause obstructive jaundice and acute haemorhagic pancreatis.

Control :

(i) Prophylaxis :

(a) Proper disposal of human faeces. (b) Avoid consumption of contaminated water, food and raw vegetables. (c) Treatment of parasitised (infected) individuals. (d) Health education to childrens in schools on sanitary laws and hygiene.

(ii) Treatment :

The infected persons should be given *piperazine citrate syrup, hexylresorcinol, santonin, mebendazole tablets* or anti-helminth drugs like oil of *chenopodium, hetrazan, tetramisole* and *dithiazanine.*

✶✶✶

Practical **2**...
Museum Study of Phylum Anelida : Neries, Earthworm and Leech

Aim : Museum Study of Phylum Annelida : *Neries*, **Earthworm and Leech.**

PHYLUM ANNELIDA

Term annelida is derived from two words one is Latin (*Annulus* = a ring) and a Greek (*Eidos* = form).

General Characters :
1. Annelids are bilaterally symmetrical animals with elongate vermiform body.
2. Body is metamerically segmented and ringed appearance.
3. They are mostly marine as well as freshwater, some are terrestrial living in burrows or tubes.
4. Body is covered with cuticle.
5. The animals are true coelomate and triploblastic.
6. Simple, unjointed, paired appendages are present.
7. Locomotion by setae, parapodia and suckers.
8. Respiration by body surface or gills.
9. Closed type of circularoty system.
10. Blood is red due to haemoglobin.
11. Excretory organs are nephridia.
12. Well developed ganglionated nervous system.
13. Sexes are separate or united.

There are about 7,000 known species in this phylum grouped into 4-classes.

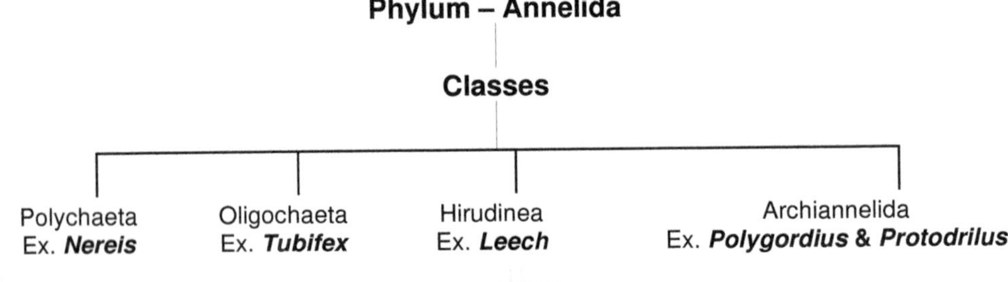

1. NEREIS

Classification :

Phylum	:	**Annelida** – Elongated, metamerically segmented, coelomate, bilaterally symmetrical, chitinous setae, excretory organs nephridia, metamerically arranged.
Class	:	**Polychaeta** – External segmentation well marked, clitellum absent, muscular parapodia with chitinous setae; sexes separate.

Order	: **Errantia** – Free living or burrowing, Accicula in the parapodium chitinous, Prostomium distinct bears appendages.
Genus	: **_Nereis_**

Habit and Habitat :

Nereis found in burrows in sand or mud of the sea shore at the tide level.

Distribution :

Cosmopolitan in distribution and found in coastal waters of Pacific and North Atlantic oceans, Europe and U.S.A.

Salient Features :

1. _Nereis_ is commonly called ragworm/clamworm.

2. Body elongated, cylindrical and divided into many metameres.

3. Body divisible in distinct head and segments (about 80).

4. The head composed of (a) _Prostomium_ with tentacles, palps and eyes and (b) _Peristomium_ carries four pairs of tentacles.

5. Each segment bears locomotory organs _Parapodia_ except the head and anal segment which posses anus.

6. On ventral side at the base of parapodia present a paired openings called _nephridiopores_ in each segment.

7. Sexes separate, gonads developed temporarily during breeding season.

8. The sexual phase of _Nereis_ is known as _Heteronereis._

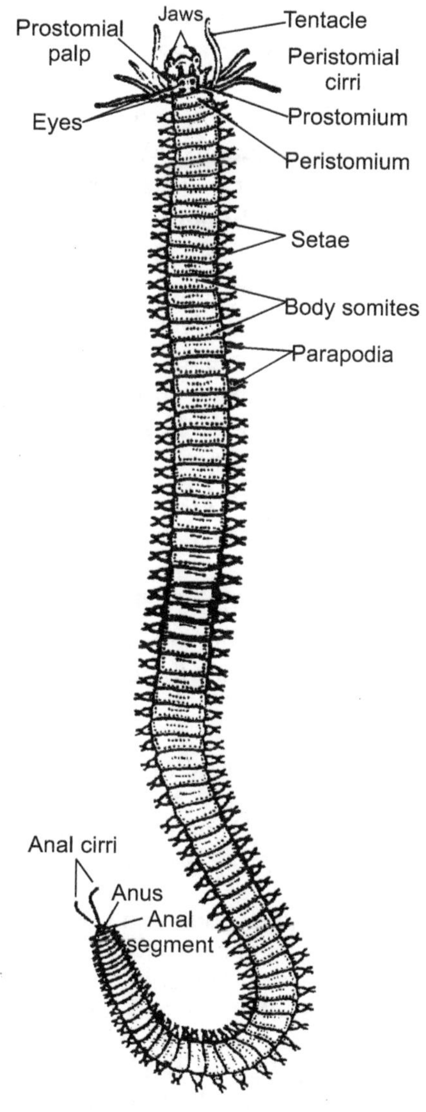

Fig. 3.2: Nereis

2. *HIRUDINARIA*

Classification :

Phylum : **Annelida** – Characters same as Tubifex.

Class : **Hirudinea** – Body segments usually - 33; No Setae, Parapodia and tentacles; Coelom is reduced by botryoidal tissue, Anterior and posterior suckers present.

Order : **Gnathobdellida** – Mouth with three toothed jaws; Each segment with five annuli except a few anterior and posterior ends; Pharynx non-protrusible, Anterior and posterior suckers well developed.

Genus : *Hirudinaria*.

Species : *granulosa*.

Habit and Habitat :

Hirudinaria is found in freshwater ponds, lakes, slow running streams and swamps. It is *sanguivorous* (blood sucking) in habit.

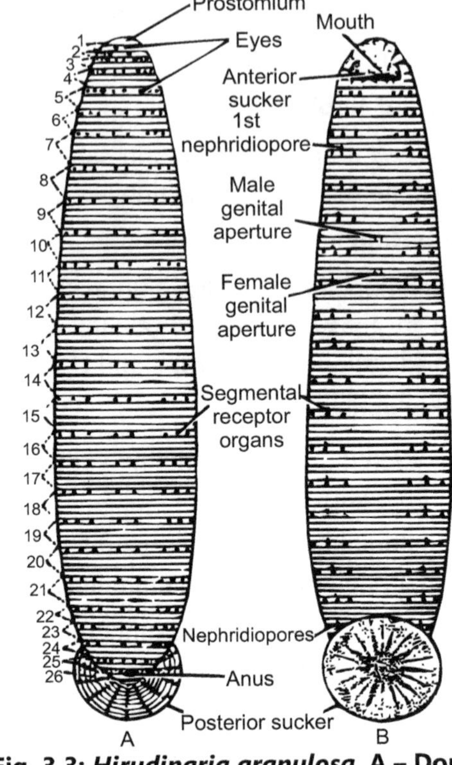

Fig. 3.3: *Hirudinaria granulosa*. A – Dorsal view; B – Ventral view

Distribution :

Cosmopolitan in distribution specially found in India and Myanmar.

Salient Features :

1. *Hirudinaria* is commonly known as *Indian cattle leech.*
2. Body soft, enlongated, vermiform, dorsoventrally flattened; about 30-35 cm in length.
3. Integument rough and verrucose, especially at the ends.
4. Body is metamerically divided into 33 somites or segments.
5. Anterior and posterior suckers are well developed.
6. Number of annuli 2 in 3, 4, 5 and 3 in 6 segments.
7. Rest of the segments with 5 annuli each.
8. Dorsally first five segments bear five pairs of eyes.
9. Triradiate mouth in the ventral anterior sucker.
10. Posterior sucker round and highly muscular.
11. Nephridiopores-17 pairs, a pair on the ventral surface of last annulus from 6 to 22 segments.
12. Hermaphrodite.
13. Reproduction sexually.
14. Leeches are the temporary ectoparasites which suck the blood of vertebrate animals.

3. *EARTHWORM*

Classification :

Kingdom	:	**Animalia**
	:	(Heterotrophic multicellular eukaryotes, animals)
Sub-kingdom	:	**Metazoa** (Multicellular, animals)
Grade	:	**Coelomata** or **Eucoelomata**
		(A conspicuous body cavity is present and is lined with coelomic epithelium)
Phylum	:	**Annelida** (segmented worms)
Class	:	**Oligochaeta**
		(Freshwater, terrestrial worms, setae are present in the skin)
Order	:	**Terricolae** (Terrestrial worms)
Genus	:	*Pheretima*
Species	:	*posthuma*

Habit and Habitat :

Pheretima is popularly known as earthworm. Earthworm is burrowing annelid found in moist soil. They are found in large numbers during monsoon.

Salient Features :

1. A mature worm measures about 150 mm in length and 3-5 mm in diameter. The anterior end is pointed and the posterior end is blunt. The body is covered by a smooth and slimy skin which is divided into many ring-like segments called *metamers*, about 80-120.

2. The body is dark brown coloured, the dorsal surface is darker than the ventral.

3. The first segment of the body is known as *Periostomium* which bears a crescentic mouth opening on the ventral surface. *Prostomium* is a fleshy lobe and is an anterior outgrowth of the periostomium.

4. In mature earthworm a thick, brown, circular band of glandular tissue, termed *clitellum* surrounds the segments 14th, 15th and 16th. The clitellum is very distinct during the breeding season.

5. A mid-dorsal line marks the dorsal side, while various reproductive openings are on the ventral side.

6. The female *genital aperture* is a single mid-ventral opening on the 14th segment.

7. The male *genital apertures* are a pair of openings on the ventral side of the 18th segment.

8. There are four pairs of elliptical openings lying ventrolaterally in the intersegmental grooves of 5/6, 6/7, 7/8 and 8/9 termed *spermathecal openings*.

9. There are two pairs of genital papillae lying on the 17th and 19th segments.

10. Dorsal pores are found on the mid-dorsal line starting from the intersegmental grooves between 12/13 onwards except the last segment.

11. The anus is found on last segment.

12. Nephridiopores are microscopic-openings on the body wall of the segment 7 onwards.

13. In all the segments, except the first, the last and the clitellar region, there are present minute bristle like chitinous structures called setae, which are embedded in setigerous sacs in the skin.

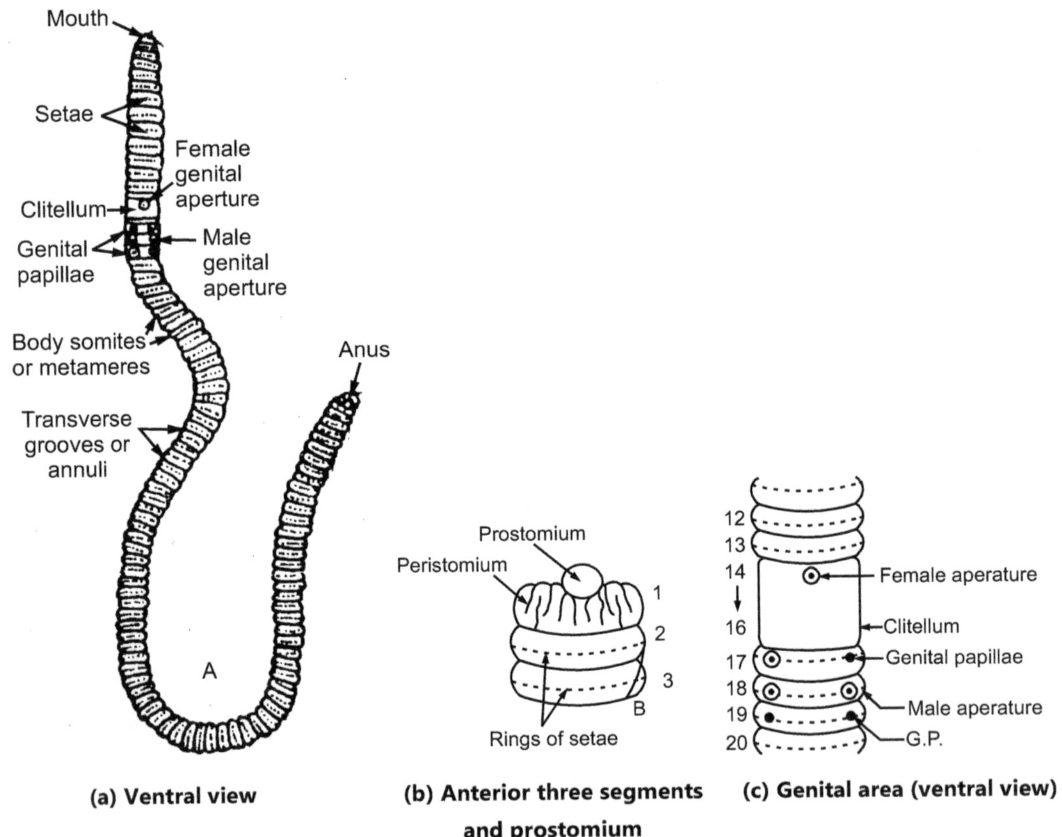

(a) Ventral view   (b) Anterior three segments and prostomium   (c) Genital area (ventral view)

Fig. 3.4: External characters of Earthworm

***

Practical **3**...

Museum Study of Phylum Arthropoda : Prawn, Crab, Cockroach, Centipede, Millipede

Aim : Museum Study of Phylum Arthropoda.
   (1) Prawn, (2) Crab, (3) Cockroach, (4) Centipede, (5) Millipede.

PHYLUM – ARTHROPODA

Arthropoda means 'jointed legs' (Greek; *arthros* = jointed; *Podos* = foot). Therefore, jointed legs is the most important characteristic structure of this phylum.

1.1 General Characters

(1)   The animals are bilaterally symmetrical and metamerically segmented.
(2)   The body is covered externally by thick, tough, non-living, organic chitinous exoskeleton.
(3)   Exoskeleton is cast off periodically and this process is called ecdysis or moulting. It is essential for the growth of the animal.
(4)   The appendages are jointed and they are modified as gills, jaws, legs etc. for various functions.
(5)   They are triploblastic animals but true coelom is reduced in adult and represented by excretory and reproductive organs.
(6)   Respiration may occur throughout the body surface but generally takes place by special structures like gills, tracheae and book lungs.
(7)   Circulatory system is of open type i.e. the blood flows in spaces or sinuses comprising haemocoel instead of blood vessels.
(8)   True nephridia are absent. Excretion occurs by green glands or by malpighian tubules.
(9)   Nervous system is of annelidan type.
(10)   Compound eyes are present.
(11)   Sexes are usually separate and often there is distinct sexual dimorphism.
(12)   Fertilization is generally internal. Development includes larval forms which undergo varying degree of metamorphosis to become adults. Parental care is often well marked.

1.2 Phylum Arthropoda is divided into Six Classes

1.2.1 Class - Onychophora

(1)   These are terrestrial, air-breathing primitive worm like animals.
(2)   They possess claws (*Onychos* = claw and *Phoros* = bearing).
(3)   The animal possesses a single pair of antennae, eyes and jaws.
(4)   Many pairs of stumpy legs are present which are not joined.
(5)   Body shows no distinct external segmentation.
(6)   ***Peripatus*** the representative of this sub-phylum is the connecting link between phylum Annelida and Arthropoda.

Example: *Peripatus.*

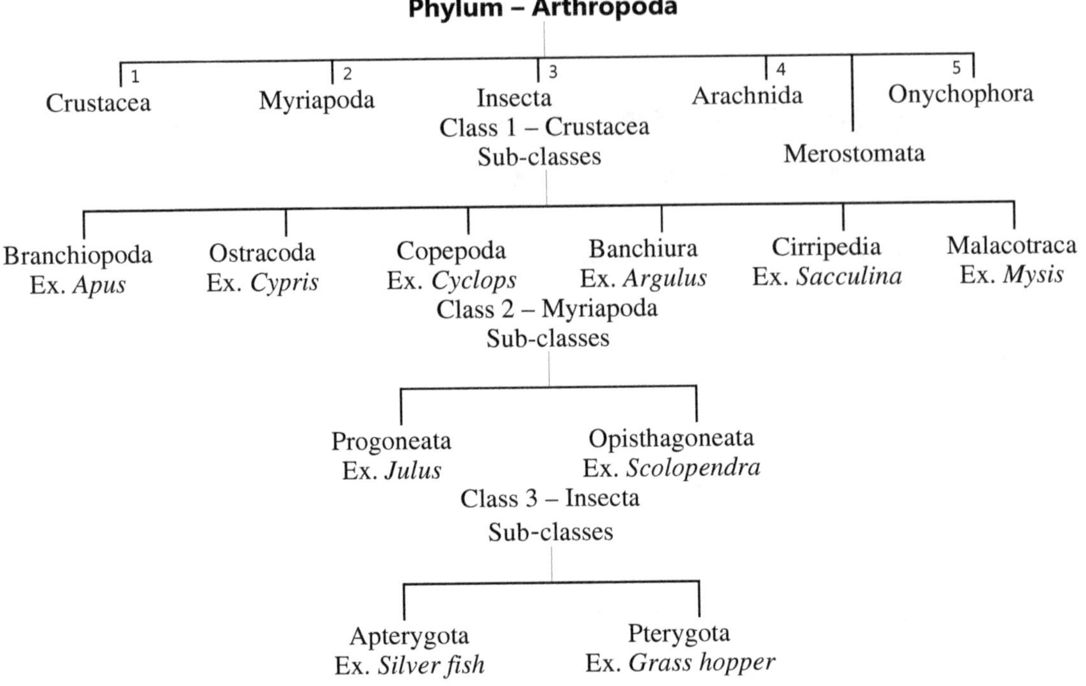

Subpylum-Chelicerata

(1) These animals lack both antennae and mandibles.

(2) Body is usually divided into two parts: anterior unsegmented prosoma or cephalothorax and posterior segmented opisthosoma or abdomen.

(3) Prosoma has six pairs of appendages, one pair of chelicerae, one pair of pedipalpi and four pairs of legs.

(4) Animals show respiration by gills, book lungs or tracheae.

(5) Coxal glands or malpighian tubules are the excretory organs.

(6) Sexes are separate. Males are smaller than females.

(7) Animals are terrestrial and predaceous.

This subphylum includes two important classes: **Merostomata** and **Arachnida**.

1.2.2 Class-Merostomata

(1) They are aquatic and marine chelicerates, with branchial respiration.

(2) Body shows two parts: anterior cephalothorax or prosoma and posterior abdomen or opisthosoma. Abdomen is again divisible into meso and metasoma.

(3) The cephalothorax is covered by a broad, horse-shoe shaped unsegmented carapace and dorsally it bears pair of lateral compound eyes and pair of median simple eyes.

(4) On ventral side cephalothorax carries one pair of chelicerae and five pairs of similar joined walking legs.

(5) Antennae are absent.

(6) Mesosoma bears six pairs of appendages of which first pair forms genital operculum and remaining five pairs of appendages bear book-gills.

(7) Metasoma is reduced, without any appendages but possesses spike like telson.

(8) Malpighian tubules are absent but coxal glands are excretory organs.

(9) Sexes are separate. Fertilization is external and development includes trillobite larva.

Example: *Limulus* (king crab).

1.2.3 Class-Arachnida

(1) Mostly terrestrial, solitary, air-breathing arthropods.

(2) Free living, predaceous or parasitic.

(3) No head, body divisible into prosoma (cephalothorax) and opisthosoma (abdomen).

(4) The cephalothorax bears simple and sessile eyes and six pairs of joined appendages. Of these one pair is of chelicerae, one pair pedipalpi and four pairs of walking legs.

(5) Antennae and mandibles are absent.

(6) Abdomen is without appendages.

(7) Cuticle is with sensory hairs or scales.

(8) Mouth parts and digestive tract are adapted for sucking.

(9) Respiration by tracheae or book-lungs or both. In aquatic forms it cutaneous or by book gills.

(10) Excretion takes place by Malpighian tubules or coxal glands or both.

(11) Sexes are separate and sexual dimorphism is inconspicuous. Males are smaller than females.

(12) Fertilization is internal. Oviparous and few viviparous. Development is direct.

Examples: *Palamnaeus, Buthus* (Scorpions), *Spiders, Ticks and Mites.*

Subphylum-Pantopoda (Pycnogonida):

(1) Aquatic, marine, small spider like animals.

(2) Body has single cephalic somite and 3 or 4 trunk somites and abdomen is rudimentary.

(3) They bear one pair of chelicere, one pair of palps, one pair of ovigerous legs and 4 to 6 pairs of long walking legs.

(4) Long porboscis bears mouth.

(5) Two pairs of eyes are present.

Examples: Sea spiders (*Nymphon, Pycnogonum*).

Subphylum-Mandibulata:

(1) The animals are terrestrial: freshwater or marine.

(2) Body is divisible either in two parts (cephalothorax and abdomen) or three parts (head, trunk and abdomen).

(3) One or two pairs of antennae, one or two pairs of maxillae and three or more pairs of walking legs are present.

(4) Gills or tracheae are respiratory organs.

(5) Malpighian tubules and coxal glands are excretory organs.

(6) Sexes are usually separate and development includes larval stage.

Subphylum mandibulata includes three important classes:

Crustacea, Myriapoda and Insecta.

1.2.4 Class - Crustacea

(1)　They are aquatic, marine and freshwater.

(2)　Some are parasitic or sedentary.

(3)　They have variously shaped bodies with three divisions; head, thorax and abdomen.

(4)　Exoskeleton is made up of thick protective, chitinous cuticle strengthened by impregnation with calcium salts, hence crustacea.

(5)　The thorax may fuse partly or wholly with head to form cephalothorax, which is partly or entirely covered by exoskeletal shield called carapace from dorsal side.

(6)　Head bears pair of stalked compound eyes.

(7)　Jointed appendages vary in number but there are two pairs of antennae, one pair of mandibles and two pairs of maxillae on head and one pair of appendages on each segment of thorax and abdomen.

(8)　First pair of antennae are uniramous but remaining all appendages are biramous. The appendages are variously modified as jaws, legs, fins, gills or accessory reproductive organs.

(9)　Respiration by gills or pseudotracheae or by body surface.

(10)　Excretion by green glands malpighian tubules are absent.

(11)　Mostly, unisexual except cirripedia. Sexual dimorphism is common. Generally, females carry the eggs.

(12)　Some groups like Brachiopoda and ostracoda show Parthenogenesis.

(13)　Development includes more or less of metamorphosis and is accompanied by free living larva nauplius.

Examples: *Lepas, Daphnia, Cypris, Cyclops, Palaemon, Lobster, Crab.*

1.2.5 Class-Myriapoda

(1)　They are terrestrial, air breathing.

(2)　Body elongated vermiform with distinct head and trunk has number of segments.

(3)　Many jointed pair of antennae are present, pair of mandibles and maxillae are present.

(4)　The body is segmented and bears one or two pairs of jointed legs.

(5)　Respiration by spiracles and tracheae.

(6)　Malpighian tubules are excretory organs.

(7)　Sexes are separate, single gonad with paired gonoducts.

Examples: *Spirobolus* (millipede), *Scolopendra* (Centipede).

1.2.6 Class-Insecta

(1)　Mostly terrestrial, aerial, air breathing animal. Rarely aquatic.

(2)　Body is divided into three distinct regions - head, thorax and abdomen.

(3)　The head is formed by fusion of six segments and bears a pair of antennae, a pair of compound eyes, few simple eyes or ocelli, a pair of mandibles, two pairs of maxillae.

(4)　The mouth parts are variously adapted for chewing, biting, piercing, sucking, lapping or siphoning etc.

(5)　The thorax consists of three segments and bears three pairs of jointed legs (hence hexapoda), two or one pair of wings present dorso-laterally. Some insects are wingless.

(6)   The abdomen consists of 7 to 11 segments and lacks appendages but genitalia are present at the end.

(7)   Salivary glands are present but liver is absent.

(8)   Respiration by tracheae opening out by paired segmental spiracles.

(9)   Heart is tubular muscular and 8 chambered situated in the abdomen

(10) Excretion occurs by two to many malpighian tubules attached to the anterior end of the hind gut.

(11) Nervous system is of the annelidan type.

(12) Sense organs are well developed.

(13) Sexes are separate, sexual dimorphism is distinct.

(14) Gonads are composed of numerous tubules or follicles.

(15) Fertilization is internal. Development usually occurs with some sort of metamorphosis. Parthenogenesis takes place in some forms.

Examples: *Mantis, Gryllotalpa*, Dragonfly, Grass hopper, *Pediculus* (Louse), Butterfly, Moth, Honey bee, Beetle.

1. PRAWN

Systematic Position:

Phylum	–	**Arthropoda**
Class	–	**Crustacea**
Order	–	**Decapoda**
Family	–	**Palaemonidae**
Genus	–	***Palaemon* or *Macrobranchium***
Species	–	***rosenbergii***

Habit and Habitat :

P. rosenbergii is freshwater inhibitant found in streams, rivers, ponds and lakes in Central and South India. It is benthic animal and omnivorous feeding algae, organic matter, insect larvae and small insects.

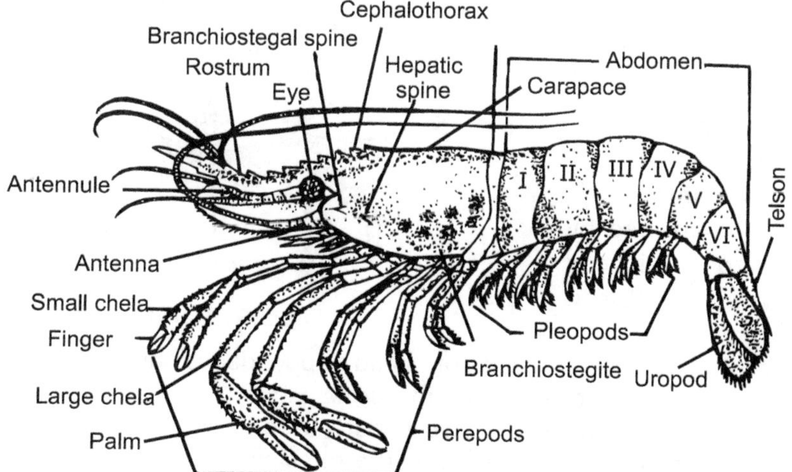

Fig. 3.5: Prawn (*Macrobrachium rosenbergii*)

Salient Features :

(1)   *Palaemon rosenbergii* is commonly called as freshwater prawn.

(2) The body is elongated, spindle shaped and about 25-40 cm long.

(3) The body is bluish green with brown or orange red patches.

(4) The body is divided into distinct two parts – an anterior *cephalothorax* and a posterior abdomen. The cephalothorax is formed by fusion of head and thorax and abdomen is formed by six movable segments.

(5) A pair of compound eyes present on head.

(6) Each abdominal segment bears a pair of jointed appendages called pleopods. The second pair of walking legs is much larger than others.

(7) The terminal conical piece of body is called as tail plate or *telson*.

(8) The cephalic region carries 5-pairs of appendages namely antennules, antennae, mandibles, maxillulae and maxillae.

(9) Prawn and prawn products of India have much appreciation as food in international market, as it is highly nutritive, tasty and palatable.

2. CRAB

Classification :

Phylum	–	**Arthropoda**
Class	–	**Crustacea**
Order	–	**Decapoda**
Sub-order	–	**Brachyura**
Family	–	**Calippidae/Portunidae/Grapsidae**
Type	–	***Edible crabs***

Habit and Habitats :

Crabs occur in freshwater, marine, brackish water habitats. They generally shows aquatic respiration by gills. Crab is carnivorous, feeding copepod, shrimps and small fishes.

Salient Features :

(1) Cephalothorax is large and covered by a hard chitinous partly calcified *carapace.*

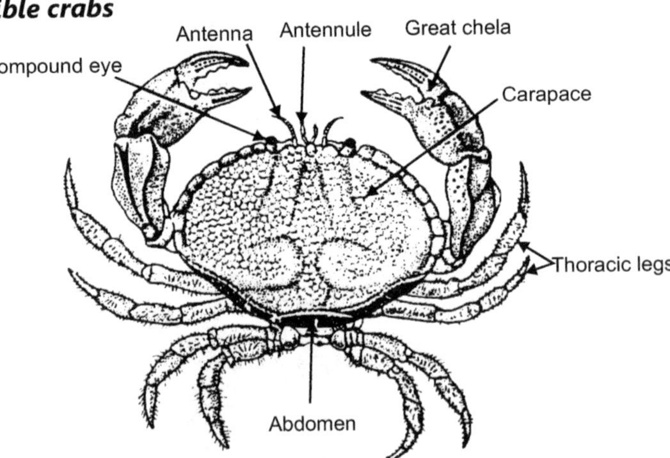

Fig. 3.6 : Marine Crabs

(2) Cephalothorax has 5-pairs of head appendages and 8-pairs of thoracic appendages. The last 5-pairs are the legs.

(3) First pair of leg are poweful called chelate, used for capturing food. They also serve as organs of offense and defence. The remaining organs are used for swimming and walking.

(4) Swimming organs are Oar-like for propulsion in water.

(5) The gills are in several series in branchial chamber covered by carapace.

(6) The abdomen is much abbreviated and kept flexed against the mid-ventral surface of the thoracic region.

(7) Abdomen of males is narrow with 2-pairs of uniramous appendages helpful in reproduction.

(8) The abdomen of female broad with 4-pairs of biramous appendages for carrying the eggs in berried ones.

(9) Life cycle have two larval stages, the *zoea* and the *megalopa*.

3. PERIPLANETA (COCKROACH)

Phylum	– **Arthropoda**	–	Jointed appendages; metamerically segmented; body cavity is haemocoel; triploblastic.
Class	– **Insecta**	–	Body is divisble into head, thorax and abdomen; three pairs of legs and two pairs of wings.
Subclass	– **Pterygota**	–	Wing usually present; abdomen devoid of appendages except genitalia and cerci; metamorphosis simple or complex.
Division	– **Exopterygota**	–	Young stages are known as nymphs; wings develop externally metamorphosis primitive or simple.
Order	– **Dictyoptera**	–	Antennae filiform with numerous segments; mouth parts are mandibulate type, forewings are lethery and hindwings are membranous; eggs contained in ootheca.
Family	– **Blattidae**	–	It includes about 35,000 species; ootheca divided into two rows of packets by longitudinal partition; similar legs with large flat coxae.
Genus	– ***Periplaneta***	–	Presence of wings in both the sexes.
Species	– ***americana***	–	Colour is light and size is large about 2.5 - 5 cm.

Salient Features :

1. Body is divisible into head, thorax and abdomen.
2. Three pairs of legs, therefore known as hexapoda.
3. Two pairs of wings, forewings are leathery and hindwings are membranous. Forewings form wing covers called *tegmina*.
4. Head bears a pair of compound eyes.
5. Spiracles and tracheae are the respiratory organs.
6. Antennae almost invariably filiform with numerous segments.
7. Mouthparts are of the chewing or mandibulate type.
8. Metamorphosis is gradual, without a pupal stage. Young stages are called nymphs.
9. Cerci many segmented.
10. Gizzard with powerful masticatory armature.
11. Head nearly or completely covered from above by large, shield like pronotum.
12. Eggs contained in ootheca.
13. *Periplaneta* is common household pest, and commonly called as ship cockroach or American cockroach.
14. Cockroach is nocturnal animal i.e. it comes out to feed at night.
15. It is omnivorous or scavengerous in its diet.
16. The adult cockroach is narrow, elongated, dorso-ventrally compressed and bilaterally symmetrical. It measures about from 30-40 mm in length and about 10-12 mm in width.

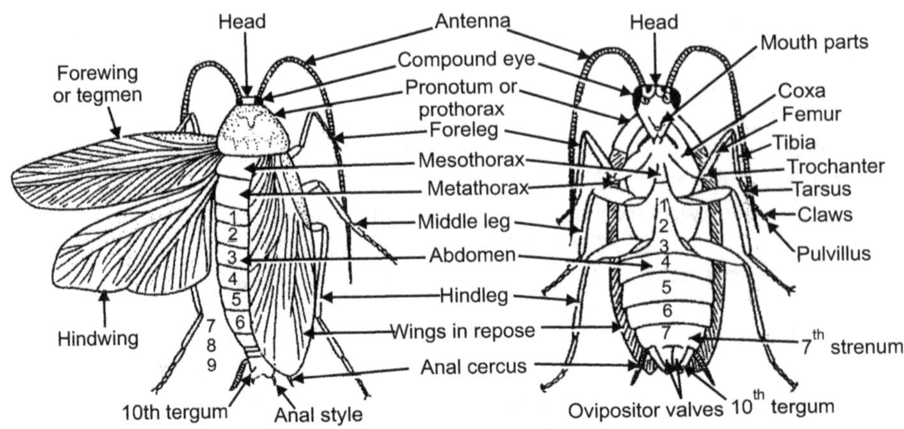

Fig. 3.7: *Periplaneta*

4. *SCOLOPENDRA* (CENTIPEDE)

Classification:

Phylum   – **Arthropoda**   : Triploblastic, body metamerically segmented, jointed appendages, body cavity is called haemocoel.

Class   – **Myriapoda**   : Terrestrial; air breathing; elongated body with many segments each bearing one or two pairs of legs.

Subclass   – **Opisthogoneta** : Genital openings are situated at the posterior end of the body.

Order   – **Chilopoda**   : Numerous trunk segments, each bearing a single pair of legs; poison jaws present.

Genus   – ***Scolopendra***

Salient Features :

(1) It is also commonly called centipede.

(2) Body elongated, dorsoventraly flattened with many segments.

(3) Body is divisible into head and trunk.

(4) Head bears a pair of antennae, a pair of mandibles and two pairs of maxillae.

(5) There are 22 segments in trunk which are alike.

(6) Each segment from 2-22 possess one pair of walking legs.

(7) The first pair of legs curved, clawed and act as poison glands.

(8) Each legs has 7 segments, that last one ending in a single claw.

(9) Anus in the last body segment.

(10) It is carnivorous, feeds on insects, spiders, worms, slugs, etc.

(11) It occurs under stones, rotten legs and in houses in damp places.

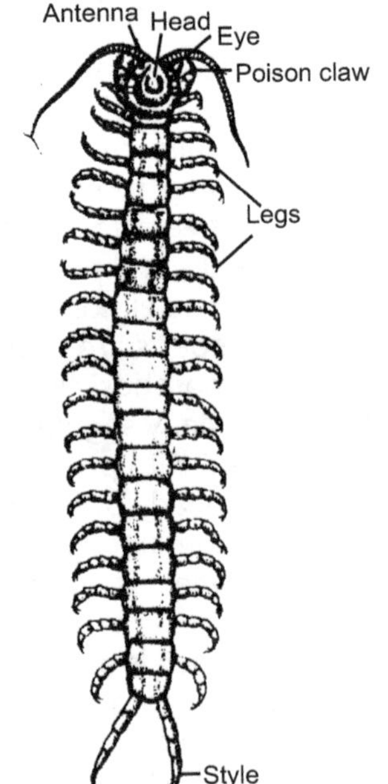

Fig. 3.8: *Scolopendra subspinipes*

5. MILLIPEDE (*JULUS*)

Phylum	–	**Arthopoda**	: Triploblastic, body metamerically segmented, jointed appendages, body cavity is called haemocoel.
Class	–	**Myriapoda**	: Terrestrial, air breathing, elongated body with many segments each bearing one or two pairs of appendages.
Subclass	–	**progeneta**	: Genital openings are situated at the anterior of the body.
Order	–	**Diplopoda**	
Genus	–	*Julus*	

Characters :

1. Body is cylindrical and segmented each segment bears two pairs of legs.
2. It is found in dark and damp places under stones and decaying leaves.
3. When disturbed usually role up into tight coil and emits an unpleasing odour from segmental scent glands.
4. Cosmopolitan distribution and commonly called wire worm.
5. Body is differentiated into head and trunk.
6. Head is small, antennae short generally 7 segmented.
7. One pair of maxillae and mandibles forming gnatha chilarium. Eyes are minute.
8. It has many legs hence called millipede.
9. Poison gland and poison claws are absent.
10. Respiration by trachea.
11. Sexes are separate.
12. Genital openings are situated at the anterior end of the body.
13. In male one or both pairs of legs of 7th segment are modified to form copulatory organs.
14. It is herbivorous and even scavenger.

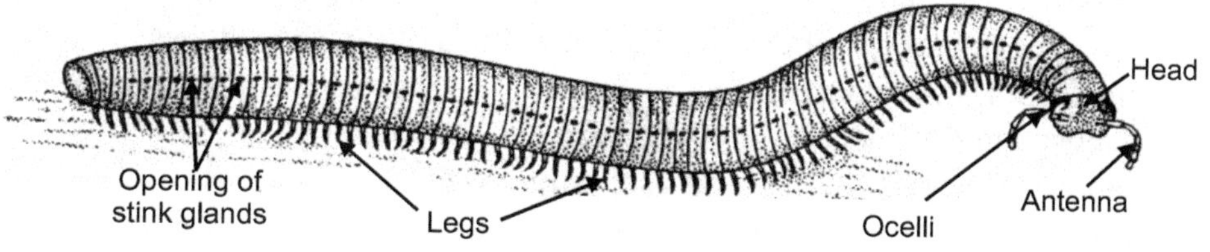

Fig. 3.9: Millipede (*Julus*)

***

Practical **4**...

Museum Study of Phylum Mollusca : Pila, Chiton, Bivalve, Octopus

Aim : Museum Study of Phylum Mollusca : Pila, Chiton, Bivalve, Octopus.

PHYLUM-MOLLUSCA

Mollusca means soft bodied animals (Latin, *molluscs* = soft). Aristotle first used this term for cuttle-fish. In latin soft nut enclosed in a thin shell is called mollusca. Thus, it is referring to the bivalve shell and soft bodied animal enclosed in the shell.

General Characters :

(1)　Molluscs are aquatic, mostly marine, some are freshwater and some terrestrial.

(2)　The symmetry is bilateral, however, gastropods and cephalopods lose their bilateral symmetry and become asymmetrical due to torsion or spiral twisting.

(3)　Body of the molluscs is soft, differentiated into four parts anterior head, dorsal visceral mass, ventral foot and mantle.

(4)　Epidermis is single layered, generally ciliated with mucous glands.

(5)　Muscular foot is present on ventral side which is locomotory organ and modified for creeping, swimming and burrowing.

(6)　A thin, muscular, fleshy fold covers the dorsal body wall called mantle or pallium. The space enclosed by mantle is called mantle cavity.

(7)　Shell is secreted by outer surface of mantle. Shell is hard, calcarious, and it may be bivalved, univalved, spiral or cone like, internal or external, reduced or even absent in some animals.

(8)　Respiration by gills called ctenidia. Body surface, mantle or lungs are respiratory organs in terrestrial forms.

(9)　Digestive system is complete. Buccal cavity contains a grasping organ, the radula with transverse rows of teeth.

(10)　Circulatory system is of open type.

(11)　Excretion is brought about by one or two pairs of sac-like kidneys.

(12)　Nervous system consists of paired ganglia, cerebral, pleural, pedal, and visceral ganglia inter connected by commissures and connectives.

(13)　Tentacles, eyes, statocysts and osphradia are the sense organs.

(14)　Sexes are usually separate, some are hermaphrodite, fetilization is external or internal.

(15)　Segmentation (cleavage) is spiral. Development may include glochidium or veliger larva.

The phylum Mollusca is divided into six classes: Monoplacophora, Amphineura, Gastropoda, Scaphopoda, Pelecypoda and Cephalopoda.

Class-Monoplacophora:

(1) Marine and they are called living fossils.
(2) Body has bilateral symmetry and is covered by a spoon or cup shaped shell.
(3) Head bears tentacles.
(4) The disc like foot has a flat creeping sole.
(5) The visceral mass is divided into five segments each with a pair of shell muscles, gills, auricles, nephridia and gonads.
(6) Buccal cavity contains radula.
(7) Stomach contains a crystalline style.
(8) Nervous system lacks ganglia.
(9) Sexes are separate.

Example: *Neopilina.*

Class-Amphineura:

(1) They are marine, bottom dwelling primitive molluscs.
(2) Body bilaterally symmetrical, vermiform, elongated and flattened.
(3) Head is not distinct, without tentacles and eyes.
(4) Foot is large, ventral, flattened and useful for creeping.
(5) Dorsal side of the body shows spicules or calcarious plates.
(6) Respiration by gills.
(7) Nervous system is primitive.
(8) Excretion by pair of kidneys.
(9) Sexes are separate or united.
(10) Fertilization is external and development includes a pelagic trochophore larva.

Examples: *Chiton, Neomenia, Chaetoderma.*

Class-Scaphopoda:

(1) Body elongated, worm like and bilaterally symmetrical.
(2) Shell is tubular, curved and open at both the ends.
(3) Foot is conical and useful for digging.
(4) Head rudimentary and without eyes and true tentacles.
(5) Long, thread-like, prehensile, knobbed processes called captacula useful for capturing the food.
(6) Ctenidia or gills are absent.
(7) Respiration by mantle.
(8) Sexes are separate and life history includes a veliger larva.

Examples: *Dentalium* (Tusk shell) and *Siphonodentalium.*

Class-Gastropoda:

(1) Gastropoda includes marine, freshwater as well as terrestrial forms.
(2) Body is asymmetrical due to torsion.
(3) Head is distinct and bears tentacles and eyes.
(4) Foot is ventral, large and flat useful for locomotion and ttachment. It bears operculum for closing the shell aperture.
(5) Shell is univalved and spiral, secreted by mantle.

(6)   Buccal cavity contains odontophore with radula having transverse rows of chitinous teeth.

(7)   Respiration by body surface or gills or lungs.

(8)   Circulatory system is open and lacunar, heart has one or two auricles and one ventricle.

(9)   The excretory system consists of a single kidney

(10)   Sexes are separate or united. Development may be direct or includes trochophore and veliger larval stages.

Examples: *Pila, Haliotis, Patella, Cypraea, Helix, Lymnea.*

Class-Pelecypoda:

(1)   Majority of animals are marine, some are freshwater and generally lead sedentary or burrowing life.

(2)   Body is bilaterally symmetrical and laterally compressed.

(3)   Head is reduced without eyes and tentacles but bears a pair of labial palps.

(4)   Foot is ventral, large, muscular wedge shaped adapted for burrowing in sand and mud.

(5)   Shell consists of two (right and left) valves movably hinged dorsally and closed by one or two adductor muscles.

(6)   Mantle comprises two (left and right) lobes which are united dorsally but hanging free ventrally and enclosing spacious mantle cavity.

(7)   Inhalent and exhalent siphons are present for the entry and exit of water into and from the mantle cavity.

(8)   Respiration by ctenidia.

(9)   Alimentary canal is coiled tube without radula.

(10)   Circulatory system consists of sinuses and vessels, heart has two auricles and one ventricle enclosed in a pericardium.

(11)   Excretion by paired kidneys.

(12)   Sexes are separate. Development includes trochophore, veliger and glochidium larvae.

Examples: *Unio, Mytilus, Pecten, Pinctada (Oyster), Solen.*

Class-Cephalopoda:

(1)   They are marine, fast swimming, highly organised, predaceous animals.

(2)   Bilaterally symmetrical body having head and trunk.

(3)   Head is prominent and bears pair of eyes and mouth.

(4)   Foot is partly modified into numerous sucker bearing arms or tentacles surrounding the mouth (hence cephalopoda). The tentacles may be 8 or 10.

(5)   The trunk is uncoiled.

(6)   Shell may be external and well-developed or internal and reduced or absent altogether.

(7)   Mouth is provided with horny jaws and radula.

(8)   Respiration by two pairs of ctenidia.

(9)   Excretion by two pairs of kidneys.

(10)   Sexes are separate.

(11)   In male one arm is hectocotylized. Serves as a copulatory organ.

(12)   Development is direct.

Examples: *Sepia, Loligo, Octopus, Nautilus.*

1. *PILA* (SNAIL)

Classification:

Phylum	–	*Mollusca*	: Body unsegmented; bilaterally symmetrical and consists of head, foot, mantle and visceral mass.
Class	–	*Gastropods*	: Body asymmetrical; spirally twisted shell in most cases it may be dextral or sinistral; torsion of visceral mass into 180°; visceral mass covered by mantle; foot large with creeping sole, kidney single.
Order	–	**Mesogastropoda**	: Presence of operculum; siphion; single auricle; single gill and monopectinate, kidney single, osphradium single.
Genus	–	*Pila*	
Species	–	*globosa*	

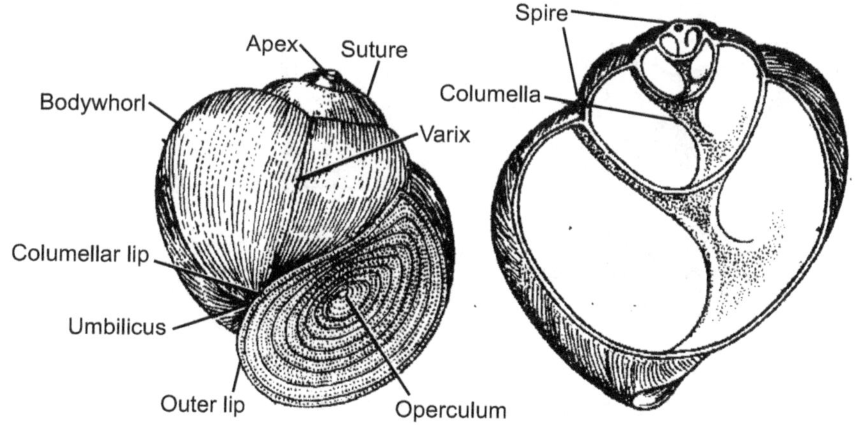

(a) An entire shell   **(b) Half of the shell ground off**

Fig. 3.10: *Pila globasa*

Salient Features:

1. Body enclosed in a thick, globular, dull yellow or brownish dextral shell.
2. Lines of growth present on the shell.
3. Head wall marked and bears a snout.
4. Tentacles are filamentous and two pairs.
5. One pair of stalked eyes.
6. Left pseudoepipodium large.
7. Foot triangular with a flat sole.
8. Visceral mass spirally twisted.
9. Respiratory organs ctenidium and pulmonary sac.
10. Shell surface is smooth and glossy.
11. Operculum calcareous.

2. *CHITON*

Classification:

Phylum	– **Mollusca**	: Body soft, unsegmented, bilaterally symmetrical and consists of head, foot, mantle and visceral mass.
Class	– **Amphineura**	: Head reduced, no eyes and tentacles, mouth and anus terminal.
Sub-class	– **Polyplacophora**	: Dorsal surface convex, ventral surface bears flat foot. 8 transverse shells.
Genus	– ***Chiton***	

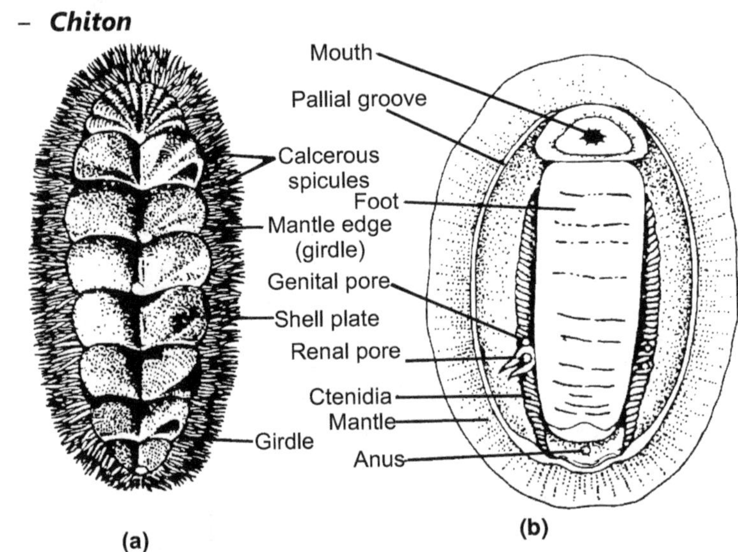

Fig. 3.11: *Chiton barnesi.* **(a) - Dorsal view; (b) - Ventral view**

Salient Features:

(1) Body is bilatreally symmetrical, elliptical, dorsoventrally flattened.

(2) Dorsal surface convex, ventral surface bears a flat foot.

(3) Eight pieces of transversely oriented shells on the dorsal surface.

(4) Head is small and not distinct. No eyes and tentacles.

(5) Foot is muscular, ventral and extending along the whole length of the body. Useful for locomotion and adhering to the substratum.

(6) Mouth and anus are terminal.

(7) Mantle covers the greater part of the body and partly covers the edges of the shell plates.

(8) Many pairs of bipectinate ctenidia (gills) are present on the either side of the body in the mantle groove.

(9) Sexes are separate and development includes trochophore larva.

(10) It is marine, sluggish animal found attached to the rock and feeds only on algae hence herbivorous.

3. *UNIO*

Classification:

Phylum – **Mollusca** : Characters same as *Aplysia*.

Class – **Pelecypoda** : Bivalved shell, body laterally compressed, head is not distinct.

Genus – *Unio*

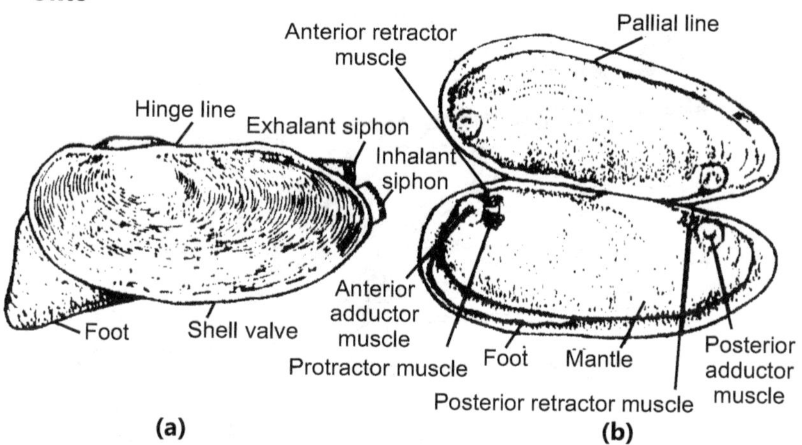

Fig. 3.12: Fresh water *mussel (Lamellidens marginalis)*
(a) An entire specimen; (b) Left valve of the shell open

Salient Features:

(1) It is also known as fresh water mussel.

(2) Body soft, bilaterally symmetrical and flattened, laterally compressed.

(3) Body is covered by bivalve shell which are equal and made up of calcium carbonate.

(4) Along the dorsal side both the valves of shell are joined together by hinge ligament.

(5) Umbo is located at the anterior end of the dorsal side.

(6) Closing and opening of the valves is done by anterior and posterior adductor muscles.

(7) Foot is large, muscular and wedge shaped useful for burrowing in sand or mud.

(8) At the posterior end of mantle inhalent and exhalent siphons are present.

(9) Two bipectinate gills, one on each side of the visceral mass are present.

(10) Sexes are separate, development includes glochidium larva.

4. *OCTOPUS* (INDIAN DEVIL FISH)

Classification:

Phylum – **Mollusca** : Body unsegmented, bilaterally symmetrical and consist of head, foot, mantle and visceral mass.

Class – **Cephalopoda** : Head well developed bears a pair of eyes; foot modified into oral arms and a siphon, mouth has a pair of horny jaws, odontophore well developed, presence of ink gland.

Order – **Octopoda**   :   Long non-retractile and oral arms 8 with sessile suckers, internal shell absent.

Genus – *Octopus*
Species – *macropus*

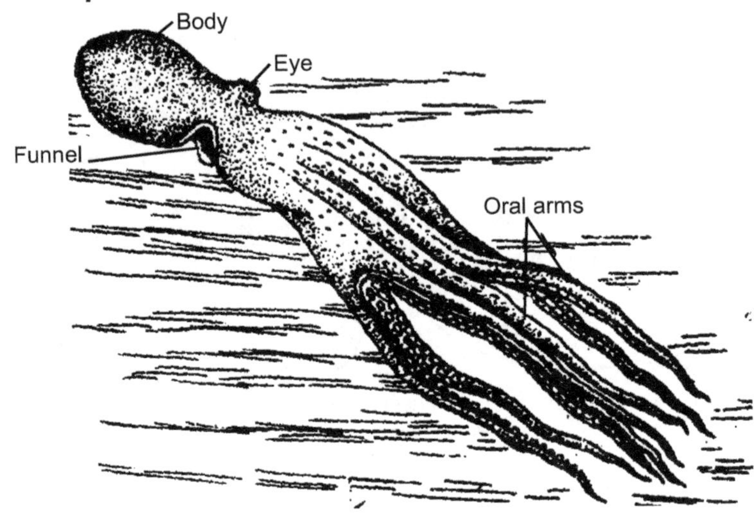

Fig. 3.13: *Octopus macropus*

Salient Features:
1. Body rounded and differentiated into head and visceral hump.
2. Head possesses 8-elongated arms with two rows of sessile, cupped sucker and a siphon.
3. Two large eyes are present.
4. Visceral mass globose, soft and fleshy.
5. Body sac-like, with conspicous white spots on reddish background.
6. Two rod-like vestiges of the shell present.
7. Female about 1.2 m and males about 1.3 m in length.
8. Males with left or right third are hectocotylized with spoon-shaped non-filamentous tip.

***

Practical 5...

Museum Study of Phylum Echinodermata : Sea Star, Sea Urchin, Brittle Star, Sea Cucumber

Aim : Museum Study of Phylum Echinodermata. Sea star, Sea urchin, Brittle star, Sea cucumber.

PHYLUM-ECHINODERMATA

The name Echinodermata literally means 'spiny skinned animals'. Greek; *echinus* = spiny, *derma* = skin. The skin or test of these animals bears prominent spines.

General Characters:

(1) Echinoderms are exclusively marine.

(2) They are gregarious, mostly free-living, slow-moving (creeping), some are pelagic and few are fixed.

(3) Animals show pentamerous radial symmetry and larvae show biradial symmetry.

(4) Body shape may be star-like, globular spherical, discoidal or elongated, flattened with oral and aboral surface.

(5) Ambulacral grooves are present on the body surface.

(6) Endoskeleton consists of closely fitted plates forming a shell or theca or test. These plates are hard and calcarious.

(7) Exoskeleton is made of movable calcarious spines.

(8) Body cavity or coelom is large, lined by ciliated peritoneum.

(9) Presence of water vascular or ambulacral system is the characteristic feature of echinoderms. Sieve plate like madreporite is present which is hydropore.

(10) Tube feet or podia are useful for locomotion, food capture and respiration.

(11) Respiration occurs through variety of structures such as papulae, (skin gills), peristomial gills, genital bursae and cloacal respiratory trees.

(12) Pedicellariae protect the delicate papulae or skin gills.

(13) Circulatory (haemal) system is reduced and lacunar. Heart is absent.

(14) Excretory system is wanting.

(15) Nervous system is primitive, lacks a brain, but consists of a circum oral ring and radial nerves.

(16) Sense organs are poorly developed and consist of statocysts, pigment eye-spots and tactile tentacles.

(17) Sexes are separate, no sexual dimorphism.

(18) Reproduction is entirely sexual, fertilization external.

(19) Development includes microscopic ciliated, transparent, free swimming larval stage.

(20) Some echinoderms reproduce asexually by transverse fission and some show autotomy by which lost body part can be regenerated.

Phylum Echinodermata is divided into two sub-phyla: Eleutherozoa and Pelmatozoa.

Subphylum-Eleutherozoa:
    (1)   They are free living without stem or stalk.
    (2)   The mouth lies on the surface facing downward and anus on aboral surface.
    (3)   Body is pentamerous.
    (4)   Tube feet or podia are locomotory as well as food gathering organs.
    (5)   Nervous system is located on oral surface.

This sub-phylum is divided into four classes: *Asteroidea, Ophiuroidea, Echinoidea* and *Holothuroidea.*

Class-Asteroidea:
    (1)   Free-living, slow-creeping and predaceous animals.
    (2)   The body is flattened, star shaped or pentagonal, radially symmetrical and differentiated into a central disc and arms.
    (3)   The arms are usually five in number but their number may vary up to 50.
    (4)   Oral and aboral surfaces are distinct.
    (5)   Endoskeleton consists of separate ossicles.
    (6)   The oral surface bears the mouth and five narrow open ambulacral grooves.
    (7)   Two to four rows of locomotory tube feet or podia are present in each ambulacral groove which are retractile and provided with terminal suckers.
    (8)   Respiration by papulae.
    (9)   Movable pincer-like spines called pedicellariae are present.
    (10)   Sexes are separate. Development includes either bipinnaria or brachiolaria larva.

    Examples: *Asterias* (Sea-star), *Pentaceros, Astropecten.*

Class-Ophiuroidea:
    (1)   Body is flattened and star shaped.
    (2)   Arms are usually five, long slender cylindrical, joined and highly flexible.
    (3)   Ambulacral grooves are absent.
    (4)   Pedicellariae, skin gills and special sense organs are absent.
    (5)   Madreporite is present on oral surface.
    (6)   Tube feet are not locomotory but useful for respiration and touch.
    (7)   Locomotion is effected by slender arms.
    (8)   Sexes are separate. Development includes free-swimming larva pluteus.

    Examples: *Ophiothrix* (Brittle star), *Astrophyton, Ophioderms, Goragonocephalus.*

Class-Echinoidea:
    (1)   The body is globular, heart-shaped or disc like, without arms.
    (2)   Endoskeleton forms a shell or corona of close fitting, immovable calcarious ossicles.
    (3)   The outer surface of shell is covered by long, movable spines and are used in locomotion.
    (4)   Outer surface shows alternative 5 ambulacral and 5 interambulacral zones or areas.
    (5)   Oral and aboral sides are distinct, mouth is on oral side and aboral surface carries the anus and madreporite

(6)   Ambulacral grooves are absent.

(7)   Tube feet protrude in two rows on each ambulacral zones having ampullae and suckers and are used locomotory, respiratory and tactile organs.

(8)   Pedicellariae are stalked and three jawed.

(9)   Mouth has five teeth with elaborate system of ossicles, all forming the characteristic Aristotle's lantern.

(10) Sexes are separate. Development includes free swimming pluteus larva.

Examples: *Echinus* (Sea urchin), *Clypeaster* (Cake-urchin), *Echinocardium* (Heart-urchul).

Class-Holothuroidea:

(1)   Body is elongated, cylindrical and without arms.

(2)   Endoskeleton is reduced to microscopic ossicles or spicules.

(3)   Skin is soft, thin or leathery without spines and pedicellariae.

(4)   Oral and aboral ends are distinct. Oral end is anterior and has the mouth surrounded by a ring of refractile, sometimes branched tentacles. These are modified tube feet and are often called buccal poidia.

(5)   Aboral end is posterior and has the anus.

(6)   Locomotory tubefeet usually present, occupying five ambulacral areas. Ambulacral grooves are absent.

(7)   Alimentary canal is long coiled and cloaca usually having respiratory tree.

(8)   Sexes are separate and development includes auricularia larva.

Examples: *Holothuria* (Sea-cucumber), *Thyone, Synapta, Stichopus*.

Subphylum-Pelmatozoa:

(1)   Mostly extinct animals.

(2)   In early life they are attached by aboral stalk supported by rows of calcarious ossicles.

(3)   Oral surface is directed upwards and bears both mouth and anus.

(4)   Tube feet are ciliated, food catching and without suckers. They function only as respiratory and tactile organs.

(5)   Spines, madreporite and pedicellariae are absent.

(6)   Main nervous system is aboral.

This subphylum includes a single living class namely, Crinoidea.

Class-Crinoidea:

(1)   Extinct or living forms, usually attached permanently or temporarily by a jointed stalk to the sea bottom. Many are free swimming and without stalk.

(2)   Body is enclosed in a cup like theca, pentamerous with upwardly directed oral surface and downward aboral surface.

(3)   The arms are long, slender usually five or ten in number branched with small alternating branches, the pinnules.

(4)   Ambulacral grooves are ciliated radiate out from mouth on oral side of arms and pinnules to their tips.

(5)   Calcarious endoskeleton is present.

(6)   Tube feet are without ampullae and suckers, and are useful for excretory, respiratory and tactile function.

(7)   Pedicellariae, spines and madreporite are absent.

(8)   Sexes are separate, development includes doliolaria larva.

(9)   Great power of regeneration is exhibited by these animals.

Examples: *Antedon* (Feather-star) *Sea-lily*.

1. ASTERIAS OR SEA STAR

Classification:

Phylum	:	**Echinodermata**
Subphylum	:	**Elentherozoa**
Class	:	**Asteroidea**
Order	:	**Forcipulata**
Genus	:	***Asterias***
Species	:	***rubens***

Salient Features:

1.   It is commonly called as 'starfish'.
2.   It is exclusively marine animal.
3.   *Asterias* is free living, crawling slowly in shallow water.
4.   Asterias is benthic animal lives on sandy bottom.
5.   It is solitary and predaceous animal.
6.   Body is flattened in oral-aboral axis and is generally pentamerous in arrangement.
7.   It has five elongated arms.
8.   Mouth is guarded by five groups of oral spines or mouth papillae.
9.   Two to four rows of locomotory tube foot or podia are present in each ambulacral groove which are retractile and provide with terminal suckers.
10.   Madreporite is present on aboral surface.
11.   Respiration is done by derminal branchiae or gills.
12.   Pedicellariae are the modified spines occur in between spines. They are useful for protection of the animal.
13.   Sexes are separate and development includes bipinnaria or brachiolaria larva.

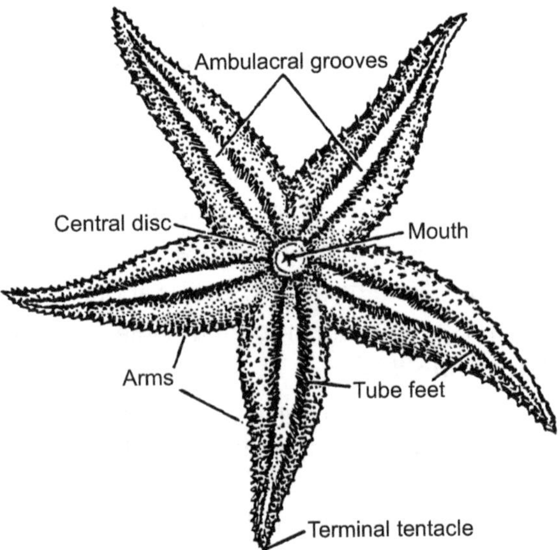

Fig. 3.14: *Asterias* or Sea Star

2. *ECHINUS*

Classification:

Phylum — **Echinodermata**

Subphylum — **Elentherozoa**

Class — **Echinoidea** — Spherical body, enclosed in shell or test, Ambulacral grooves and anus are absent. Stalked pedicellariae with three jaws.

Genus — *Echinus*

Salient Features:

(1) It is commonly known as sea *urhcin*.

(2) Body nearly spherical with flat oral and domed aboral surface.

(3) The body of the animal is enclosed in a rigid globular shell or corona which is formed by closely fitted calcarious plates.

(4) Whole surface of the test or shell except peristome and periproct covered by movable spines articulated to it.

(5) Three jawed pedicellariae and sphaeridia are present among the spines.

(6) Mouth is located in the centre of oral pole and it is surrounded by soft membrane called peristome. Through the mouth project five teeth of Aristotle's lantern.

(7) Tube feet are arranged in five double, in the ambulacral areas.

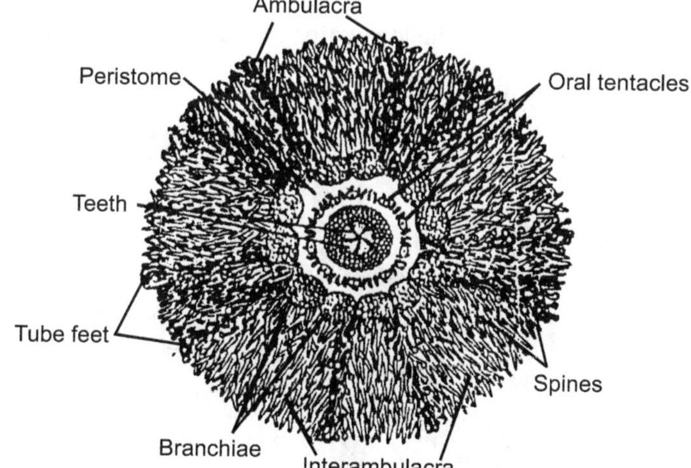

Fig. 3.15: *Echinus esculentus*. Oral view

(8) The surface of test is divided into alternating ambulacral and interambulacral areas.

(9) Sexes are separate. Development includes enchinopluteus larva.

(10) Macheporite and gonopore near the aboral anus.

(11) It is a marine animal found in the sea in the rocky places.

3. BRITTLE STAR (OPHIODERMA)

Classification:

Phylum	– **Echinodermata**	– Characters same as Star fish.
Sub-phylum	– **Elentherozoa**	
Class	– **Ophiuroidea**	– Bases of the arms distinctly marked off from the disc. No ambulacral grooves, anus and intestine. Madreporite on the oral surface, bursa usually ten.
Genus	– **Ophioderma**	

Salient Features:

1. It is also called as brittle star.
2. Presence of pentagonal disc and five long arms.
3. Oral and aboral body surfaces.
4. Pentagonal mouth located on oral surface with several oral papillae and granules.
5. Tube feet and anus are absent.
6. Madreporite is located on oral surface.
7. Arma are long, slender, cylindrical and flexible with small spines.
8. Sexes are separate.
9. It is marine animal found in shallow water to a depth of 280 metres.

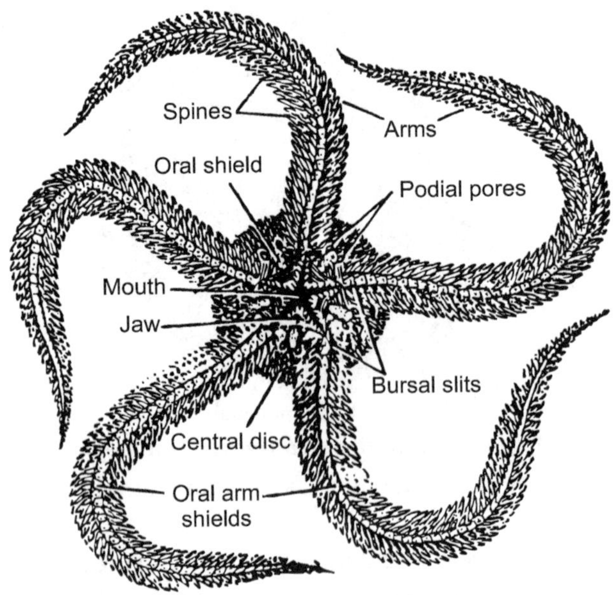

Fig. 3.16: Brittle Star

4. HOLOTHURIA

Classification:

Phylum   – **Echinodermata** – Characters same as Echinus.

Subphylum – **Elentherozoa**   – Characters same as Echinus.

Class   – **Holothuroidea** – Cylindrical body, elongated in oral-aboral axis; anus is absent, a circlet of long tentacles at the oral end.

Genus   – *Holothuria*

Salient Features:

(1) Body is cylindrical with bilateral symmetry, bearing mouth and anus at opposite ends.

(2) Skin is soft, thin and without spines and pedicellariae.

(3) Circlet of long 20 to 30 tentacles at the oral end.

(4) Body bear numerous tube feet which are locomotry on the dorsal surface.

(5) Well developed respiratory tree.

(6) Madreporite is internal.

(7) Sexes are separate.

(8) Development includes auricularia larva.

(9) The sand containing organic food is taken in the mouth with the help of tentacles.

(10) *Holothuria* is marine found in shallow tropical and subtropical waters, on the sandy bottom.

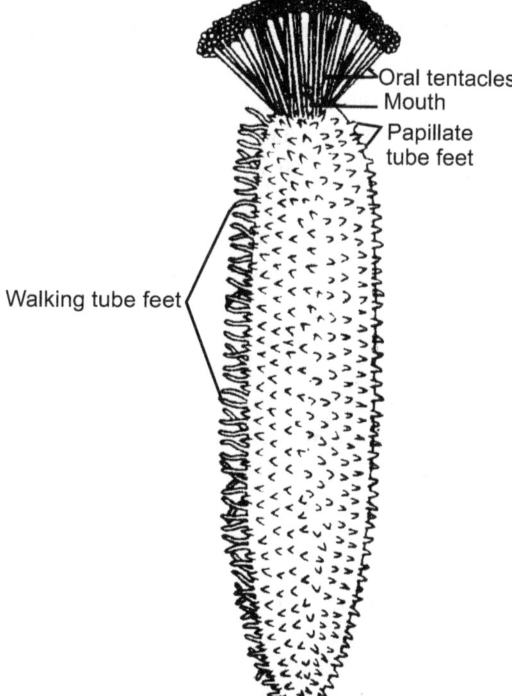

Fig. 3.17: *Holothuria nigra*

***

Practical **6**...

Study of Permanent Slide of Mouthparts of Insects : Mandibulate, Piercing and Sucking, Chewing and Lapping

Aim : To study permanent slides of mouth parts of Insects :

Mandibulate, Piercing and sucking, Chewing and lapping.

The mandibles and maxillae, along with the labrum and hypopharynx form the mouth parts, useful for feeding. Insects show diverse modes of feeding and hence mouth parts are variously modified in different insect groups to suit their modes of feeding. The study of mouth parts of insects is important to control the insect pests.

Following are the main modifications which occur in different insects :

(1)   Biting and chewing mouth parts (Cockroach).

(2)   Chewing and lapping mouth parts (Honey bee).

(3)   Piercing and sucking mouth parts (Plant bug/Bed bug and mosquito).

(4)   Sponging and lapping mouth parts (Housefly).

(5)   Siphoning mouth parts (Butterfly).

(a) Biting and Chewing Mouth Parts (Mandibulate Type):

These are the primitive mouth parts and from them all other types or mouth parts have evolved. They are also called as mandibulate type of mouth parts and are meant for pinching off, chewing up and swallowing the pieces of plant and animal tissues. Chewing type of mouth parts occur in **Cockroaches**, Grasshoppers, Crickets, Termites, Beetles, Dragon flies and Silver fish. These mouth parts consists of following parts :

(a) Labrum : It is broad flap having variable shape and size hanging from the clypeus in front of other mouth parts. It forms the roof of the mouth cavity and show slight upward and downward movement. It has on its posterior (ventral) side a membranous swollen area, the *epipharynx* which bears taste buds.

(b) Mandibles : They are the pair of mandibles which are heavily sclerotized, triangular structures lying just behind the labrum. Each mandible bears teeth like denticles, which interlock while capturing the prey. The mandibles are moved by a pair of abductor and adductor muscles. The mandibles are useful for biting and chewing the food.

(c) Maxillae : The maxillae are paired situated on the sides of the mouth behind the mandibles. They are regarded as the appendages of 5th head segment. Each maxilla consists of a basal two pieces, a proximal *cardo* and *distal stipes*. This portion is called *protopodite*. The inner endopodite has also two segments medial jaw like *lacinia* and lateral blunt *galea*. The outer exopodite or palp usually 5 segmented called *maxillary palp* functioning as a tactile organ. The lacinia and galea bear taste buds so they taste the quality of food. The maxillae are useful for holding the food material.

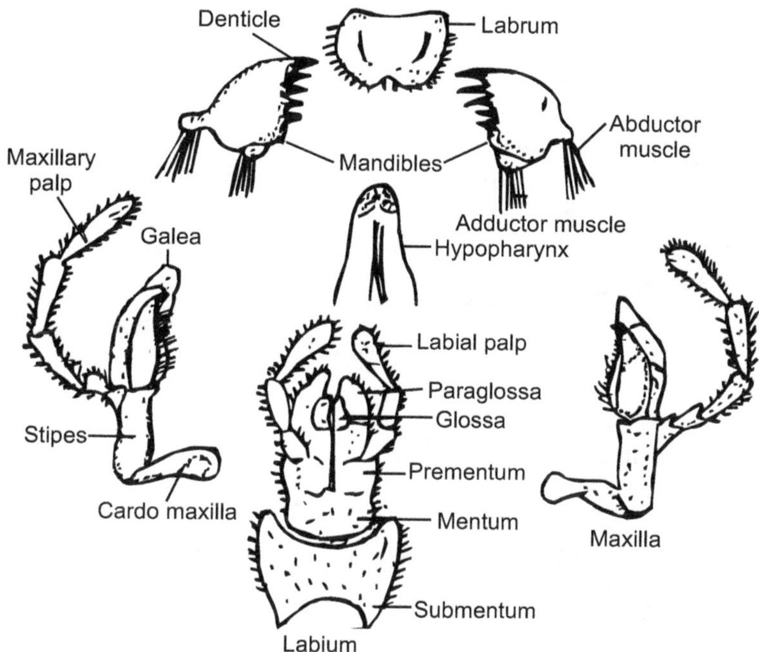

Fig. 3.18: Biting and Chewing (Mandibulate) Type of Mouth Parts

(d) Labium or Lower Lip : This is also called 2nd pair of maxillae, basically fused to form a broad plate divided by a transverse suture into 2 parts, the basal *postmentum* and the distal *prementum*. The postmentum may be divided further into a basal *submentum* and distal *mentum*. The prementum bears a pair of usually 3 jointed palps called *labial palps* laterally. In between the palp lie two *glossae* and *paraglossae*. They are commonly called *ligula*. The labial palps are sensory and help in testing the quality of food as they contain chemoreceptors. The glossae and paraglossae together prevent the loss of food particles while feeding. They also aid in pushing the masticated food into the preoral cavity. These organs are formed from the 6th head segment.

(e) Hypopharynx : It is a short, median, tongue like appendage in the mouth cavity between the labium and mandibles. It is not muscular nor glandular but supported by few narrow sclerites. The common salivary duct opens into the base of the hypopharynx.

(b) Piercing and Sucking Mouth parts:

These mouth parts occur in **bugs**, aphids, scale insects, **mosquitoes**, fleas and some flies. They are used for sucking the sap or blood by piercing into plant or animal tissue.

(a) Mouth Parts of Mosquito : These mouth parts are used for piercing and sucking the blood from host. The labrum, mandibles, maxillae and hypopharynx form long, pointed stylets for piercing the host skin. The labrum is grooved ventrally. The hypopharynx encloses the salivary duct. The labrum, hypopharynx by getting closely applied together enclose the food channel for drawing blood. The labium forms a long hairy, dorsally grooved *proboscis* tipped with a pair of sensory *labellar* lobes representing the labial palps. It serves as a sheath for the stylets at rest but folds up as the stylets pierce the host tissue at the time of feeding. The mandibles are absent in the male. Maxillary palps are short and 3 jointed.

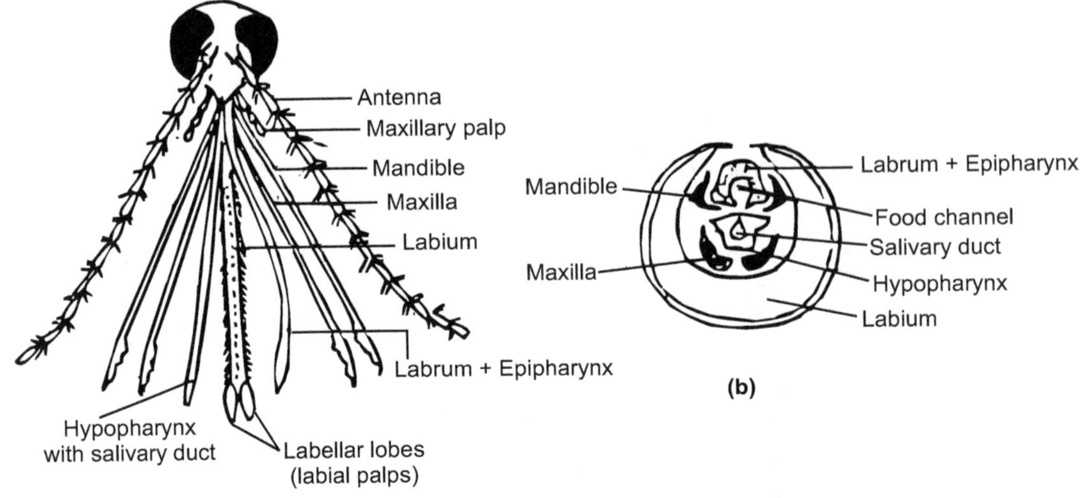

(a) Mouth Parts of Female *Culex*   **(b) T.S. of Mouth Parts of Mosquito**

Fig. 3.19: Piercing and Sucking Mouth Parts

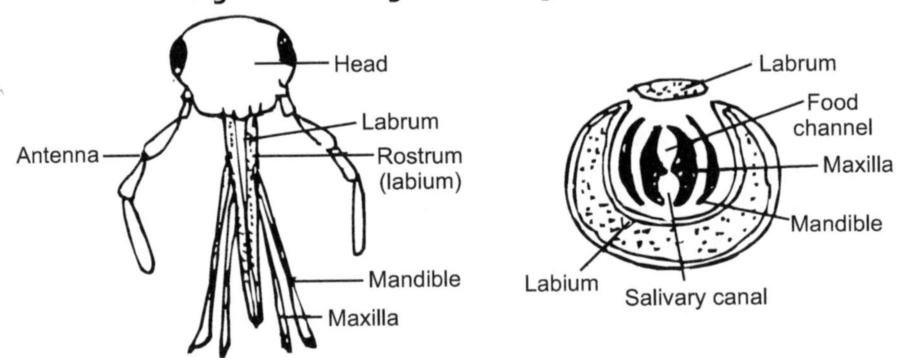

(a) Piercing and Sucking Mouth Parts of Bug   **(b) T.S. of Mouth Parts**

Fig. 3.20

(b) Mouth Parts of Bug : In the bugs, the labium forms usually a 3 *jointed rostrum* or *proboscis*. The mandibles and maxillae form four long, needle-like piercing stylets. The proboscis encloses the stylets. The hypopharynx is a short lobe within the base of the proboscis. The inner stylets in the proboscis i.e. maxillae fit together in such a way as to form the food and salivary channels. Maxillary and labial palps are absent.

(c) Chewing and Lapping Mouth parts:

These type of mouth parts are found in some Hymenoptera (**bees** and **wasps**), particularly in worker bees. They are used for collecting nectar and pollen of flower and moulding wax. These mouth parts serve for both biting and licking. The labrum lies beneath the clypeus and fleshy epipharynx projects below the labrum.The mandibles are smooth and spatulate. The worker bees uses them in building the cells of the hive. The maxillae lack laciniae have vestigial palps and bear long blade like galea. The labium has elongated palps, reduced paraglossae and elongated curved glossae united to form a retractile tongue bearing a *honey-spoon* or *labellum* at the distal free end.

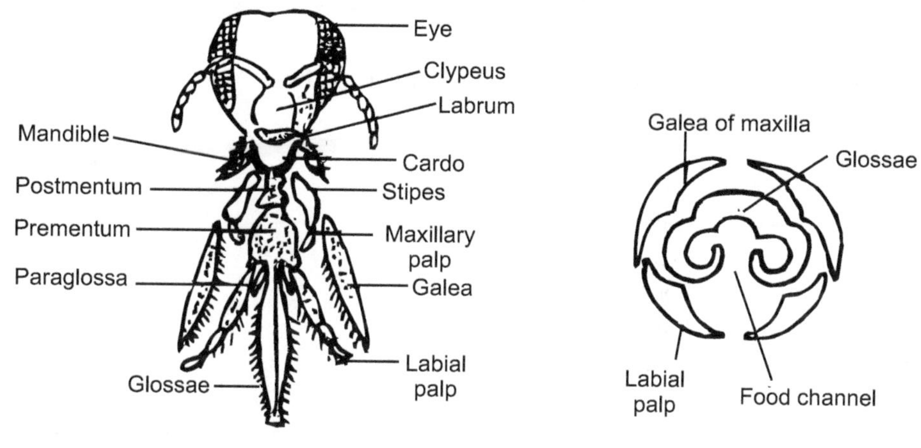

(a) Mouth Parts of Worker Honey-bee　　　　**(b) T.S. of Mouth Parts**

Fig. 3.21: Labellum (Honey Spoon)

While feeding, galeae and labial palps are brought close together forming a long hollow proboscis, which can be inserted deeply mto the corolla of a flower. Nectar is drawn up into the proboscis by back and forth movement of the tongue in it.

***

Practical **7**...

Study of Types of Shells in Mollusca

Aim : To Study Types of Shell in Mollusca. *Pila*, *Bivalve*, Chiton, Sepia.
Shell of Pila:
Classification:

Phylum	–	Mollusca
Class	–	Gastropoda
Subclass	–	Prosobranchia
Order	–	Pectinibranchiata
Genus	–	*Pila*
Species	–	*globosa*

Shell of *Pila*:

The shell of the *Pila* is somewhat globular in form about 60 mm wide and lemon-yellow, brownish or even blackish in colour. It is secreted by the mantle. The shell is made up of a single piece therefore it is called *univalved* (Fig. 3.23). The shell is also known as *unilocular* i.e. undivided internally. It is somewhat smaller in the male than in the female. The shell of *Pila* as in other Gastropoda consists of elongated, hollow cone closely coiled in a spiral manner round a central vertical hollow axis called *columella*. A single revolution of the shell around the axis is called a whorl. There are $6\frac{1}{2}$ whorls in all, and these become larger from the top to the base of the shell. The top of the shell is called the apex and it is the oldest and the first formed part of the shell from which growth of shell has proceeded below the apex is a spire consisting of several successively larger whorls or coils followed by largest whorl or body whorl which encloses greater part of the body. The whorl next above the body whorl is called penultimate whorl. All the whorls except the body whorl, are collectively referred to as the spire. The line of contact between the adjacent whorls is called the suture.

The surface of the whorls is marked by faint vertical ridges, the lines of growth. The prominent ridges or lines of growth are called varices (singular varix). Internally, all the whorls freely communicate with one another and there are no separating partitions between them. The body whorl opens out by a large lunate-oblong aperture, the mouth. The margin of the mouth is called peristome, from which the head and foot of the living animal can protrude.

The outer wall of peristome, away from the columella is called the outer lip, while the inner wall, towards the columella forms the inner or columellar lip.

The columella, or the central axis of shell is in the shape of a hollow twisted rod. It is useful for attachment of columellar muscles of the head foot complex.

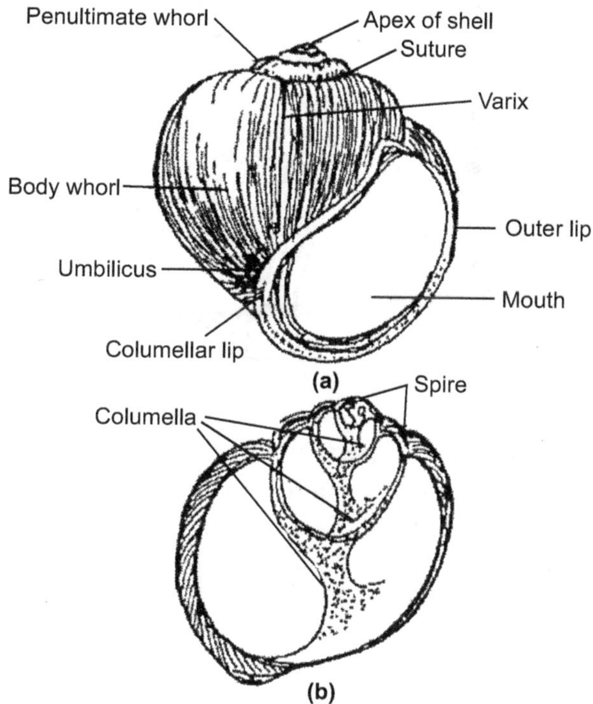

Fig. 3.22: *Pila globosa:* (a) Shell seen from ventral surface, (b) Shell seen from dorsal surface

The columella opens below to the exterior by a narrow opening adjacent to the inner lip called umbilicus. The shells with a umbilicus are called perforated or umbilicated shells, as in *Pila*. Shells lacking umbilicus are called imperforated or unumbilicated shells as in *Triton*.

Since, the shell of *Pila* is coiled in nature, therefore it has a characteristic orientation. If the shell of *Pila* is viewed from the ventral side with the peristome facing the observer, the mouth lies to the right of the columella and shell is spiralled clockwise. Such shell is called right handed or dextral. Most of the shells of *Pila* are *dextral*. Rarely, however a shell of *Pila* may coil anti-clockwise i.e. when viewed, the mouth is towards the left of the axis and such type of shell is called *sinistral*. This type of shell is abnormal, however it is a hereditary trait, which is transmitted from generation to generation like eye colour or hair colour in humans. Dextral (i.e. right handedness) character in *Pila* is dominant over sinistral (i.e. left-handedness).

Operculum:

The mouth or aperture of the shell is closed by a flat oblong calcarious plate, the operculum (Fig. 3.24). It is attached to the hinder part of the foot. It is formed by calcarious material secreted by cuticular secretion of glandular cells in the foot.

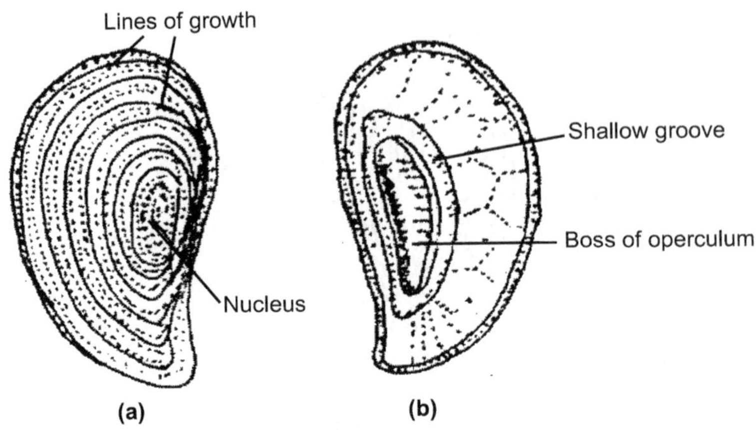

Fig. 3.23: *Pila globosa*, **Operculum (a) Outer view, (b) Inner view**

The operculum has same outline as the mouth of shell and fits tightly a little inside the peristome. Like the shell, the outer flat surface of operculum shows many concentric rings of growth and around small subcentral, the nucleus. The inner surface of operculum has an elliptical boss for attachment of muscles. The boss is cream coloured and it is surrounded by a groove. The boss serves for the attachment of muscles that operate the operculum. The operculum is useful for closing the mouth of shell when *Pila* has retracted into the shell.

Shell of Bivalve:

Classification:

Phylum	–	Mollusca
Class	–	Pelecypoda
Order	–	Eulamellibranchiata
Genus	–	*Unio* or Anodonta

Shell:

A typical bivalve shell consists of two similar, more or less oval, usually convex valves which are attached and articulate dorsally with each other. Each valve bears a dorsal protuberance called the umbo which rises above the line of articulation. Umbo is the oldest part of the shell, and the concentric lines around it are lines of shell growth.

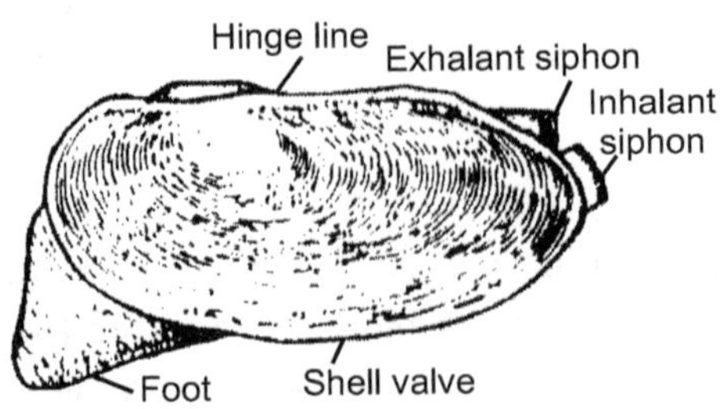

Fig. 3.24: Bivalve shell

The two valves are attached by elastic protein band called the hinge ligament, which is covered above with the *periotracum*. The hinge ligament is so constructed that when the valves are closed, the dorsal part is stretched and the ventral or inner part is compressed. Thus, when abductor muscles relax, the ligament causes the valves to open. The inner surface of each valve possesses dorsally along the hinge line, small sharp ridges and teeth like projections separated by grooves or sockets. These are called hinge teeth. The teeth of one valve fit into corresponding sockets of the other to present lateral slipping.

The valves of the shell are pulled together by two large dorsal muscles, called adductors. Near the impression of the anterior adductor muscles are two impressions. There is pallial line between anterior adductor muscles are two impressions. There is pallial line between anterior and posterior adductor muscles. It is formed due to the insertion of muscle fibres from edges of mantle.

<h2 align="center">Shell of CHITON</h2>

Classification:

Phylum	–	Mollusca
Class	–	Amphineura
Order	–	Polyplacophora
Sub-order	–	Chitonina
Family	–	Chitonidae
Genus	–	*Chiton*

Shell:

The body of the *Chiton* is elliptical, bilaterally symmetrical and dorsoventrally compressed and consists of shell, foot, mantle and the visceral mass. The shell is made up of eight transverse, overlapping calcarious plates or valves, arranged in a longitudinal row. The shell forms a solid armour covering the dorsal surface. The name of the order *polyplacophora* is derived from the nature of the shell. The shell plates are movable upon one another and allow the animal to roll up like a wood louse. The first shell plate is called *cephalic* and the last or eighth plate is called *anal* and they are hemispherical in shape. The remaining shell plates are called *intermediate* shell plates which are rectangular and often keeled middorsally. The posterior edge of each plate overlaps the anterior edge of the next behind. Each shell plate is made up of two distinct layers. The upper layer or *tegmentum* which consists of organic conchiolin matrix impregnated with calcium carbonate.

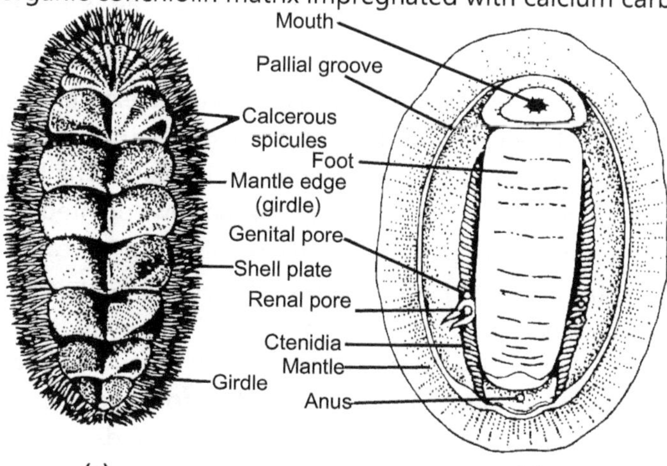

Fig. 3.25: *Chiton.* **(a) – Dorsal View, (b) – Ventral View**

This layer is perforated by numerous vertical canals containing sense organs and its upper exposed surface is sculptured. The lower thick and dense layer is called articulamentum consists of calcium carbonate only.

Foot:

The *chiton* has large, elliptical, muscular and sucker like ventral foot with a flat sole. The foot is abundantly supplied with slimy secretion and is adapted for creeping as well as clinging. *Chiton* is very sluggish animal moving by employing mucous and waves of muscular activity. If abundant food is available to one spot then the animal remains there for longer period. Foot is primarily useful for adhesion. The mantle girdle is also used for firm adhesion. When animal is disturbed, the girdle is lowered and clamped down tightly against the substratum. The inner margin of the girdle is then raised, thus creating a vacuum so the adhesion is improved. As a result, *chiton* is rarely disloged. These animals are generally found on smooth rock surfaces which provide better adhesion.

Shell of SEPIA (CUTTLE FISH)

Classification:

Phylum	–	Mollusca
Class	–	Cephalopoda
Order	–	Coleoidea
Sub-order	–	Decapoda
Family	–	Sepiidae
Genus	–	*Sepia*

It is a marine mollusc living usually in shallow coastal waters. Body is divisible into head, coller and trunk. Head bears pair of large eyes and five pairs of arms surrounding the mouth.

Shell:

In most of the molluscs, the calcarious protective shell is external, but in case of *Sepia* the shell is internal and it is enclosed in the sac of mantle on the dorsal side. The shell is secreted by the mantle. The shell is flat, broad and oval in shape. The broader and rounded oval end of the shell is called *pro-ostracum* and the narrow, pointed abroal end is called *rostrum* which is projecting into spine. The shell is made up of calcarious material and provides rigidity to the trunk like an endoskeleton. The calcarious material is arranged in fine parallel layers called laminae, enclosing spaces containing fluid and gas. Due to this arrangement the light shell is useful as a hydrostatic organ or float. It also helps in maintaining equilibrium of the body. The shell is rather soft and spongy. These shells are used as bill sharpener as well as a source of calcium to caged birds.

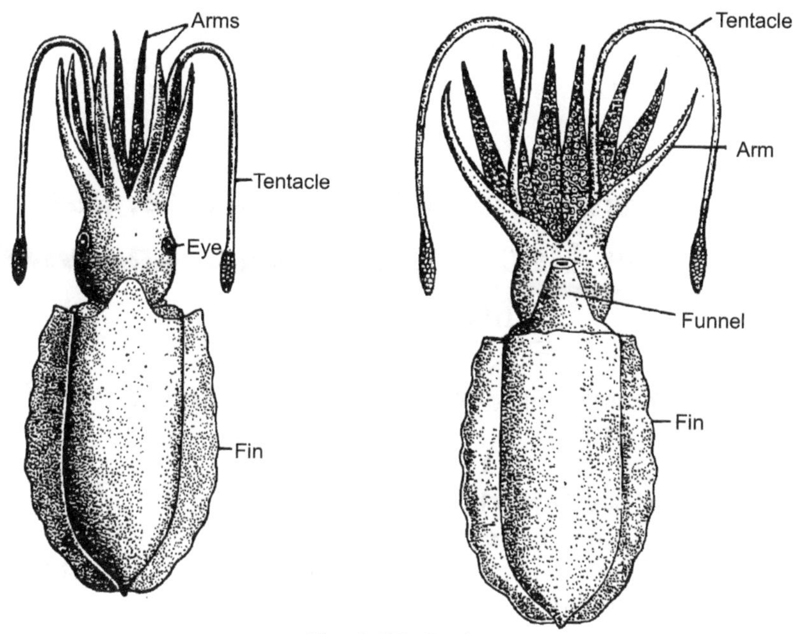

Fig. 3.26: *Sepia*

Foot:

In case of *Sepia*, the foot is modified into 10 arms, of which 8 arms are short and 2 are long. The eight smaller arms bear stalked suckers. The long two arms are called tentacles and provided with suckers only towards their free ends. The oral arms are useful for capturing the food.

***

Practical **8**...

Economic Importance of Honey Bee, Lac Insects, Silkworms, Red Cotton Bugs, *Anapheles* Mosquitoe

Aim : To Study the Economic importance of Honey bee, Lack insects, Silkworms, Red cotton bug, Anapheles mosquito.

1. HONEY BEE

Economic Importance of Bee Products:

Apiculture or beekeeping is the technique of rearing honey bees for number of products like honey, wax, pollen, propolis, bee-venom, royal jelly. Among these honey and bees wax are major products and rest are minor.

1. Honey :

Honey is made from the nectar, a sugary fluid, secreted by the nectar glands (nectaries) situated at the bases of flowers of trees, shrubs and herbs. The bees draws nectar by its tongue (Proboscis) and carries it to the hive in its crop or honey stomach; where it gets mixed with the saliva and undergo certain chemical changes due enzyme action. In the hive, it is regurgitated and deposited by the hive mates into the cells which are used for collection of nectar and its subsequent ripening into honey. The nectar is composed of 20-40% sucrose which is (reduced) converted into *dextrose* (glucose) and *levulose* (fructose) by the enzyme invertase. This enzyme is present in the nectar but more is added by the bees from their salivary glands. Ripening of honey takes place by the action of invertase enzyme and evaporation of water by fanning of wings.

Honey is very sweet in taste and white to black in colour with variable smell depending on the juices collected from different flowers.

Economic Importance and Uses of Honey :

Honey is a necessary item at puja (worship) in Hindu religion. Catholics prepare *mead* (an alcoholic drink made from honey) and in Quran there is a special chapter on uses of honey. Besides this Honey has great food and medicinal value.

   (i)   **Food value :** It is estimated that 200 gm of honey provides as much nourishment as 11.5 litre of milk or 1.6 kg cream or 330 gm meat. 2.1 gm of honey provides about 67 kcal of energy or 1 kg contains 3350 calories. The ingredients of honey like sugars, minerals, vitamins A, B and C are easily absorbed by alimentary canal. Honey is taken by healthy as well as those are ill persons. It is recommended as food for infants, the aged. It has been shown that it helps to build up haemoglobin of blood. Honey gives special flavour to foods. Such as candies, baked goods, fruits, icecreams etc.

(ii) Medicinal value : Honey is mildly laxative, antiseptic and sedative, so it is used as an ingredient in Ayurvedic and Unani systems of medicine. Just as lactose is used as carrier of Homeopathic medicines. Some tested medicinal uses of honey are its uses as laxative, blood purifier, a preventive against cold, cough and fever, curative for sores, eye ailments, ulcers on tongue, sore throat and burns. Alkaline property of honey does not produce acidosis or flatulence. The *Chavanprash* contains honey as one of the important ingredient. Its regular use is recommended after severe cases of heart attack for malnutrition, indigestion and diabetics. It is also found that *typhoid germs* are killed by honey within 48 hours, those of *branchio pneumonia* in four days and of dysentery in 50 hours.

(iii) Other uses of honey : Besides food and medicinal value, honey is used in number of ways, for example : it is used in the preparation of bread, cake, candies and biscuits due to its preservative nature. It is also used to prepare alcoholic drinks (i.e. mead). In laboratory, honey is used to stimulate the growth of plants, the bacterial culture, in insect diet and in the preparation of poison baits for fruit flies.

2. Pollen :

In the last few decades, apiculture has started to produce large quantities of pollen and royal jelly. There are high contents of biologically active substances and other valuable compounds which are of great benefit for human health. Bees begin gathering pollen immediately after their spring orientation flight and continue collecting it until late autumn which reaches its peak by summer. It is a main source of protein, fats and minerals in the honey bee diet.

Uses :

Pollen is mainly used by brood for growth. It can be fed to stimulate egg laying. Bees make bee bread from pollen. Pollen neutralizes the activity of the intestine, like colitis or chronic constipation, improves the appetite, increases haemoglobin, lowers blood pressure and is useful in anaemia. Gonadotropic hormones are found in pollen of date palm, so it is used for treating sterility. It is also suggested that pollen can maintain the health of the prostate gland in man. Chocolate coated pollen candies and other pollen products have been sold as health foods. The pollen from known plant sources is used in desensitizing persons with pollen allergies.

3. Royal Jelly :

Royal jelly is secreted by the hypopharyngeal and mandibular glands of the worker bee normally 5 to 15 days of age. It is fed to the queens throughout their larval and adult lives and also to young worker and drone larvae. It is synthesised during the digestion of pollen and is high in proteins. It can be collected from the cells of young worker bees and is present in greater quantity in the cells of queens. It is the substance which is responsible for the differences between the development of a worker and a queen. It is difficult to study the difference in the food because the chemical substances responsible for the changes are unstable and breakdown quickly during the process of collection and storage.

Uses :

It is used as food for rearing queen bees. It is particularly used in treatments of cardiovascular disorders and influenza. It also acts as a diuretic agent. But its medicinal and food value is not established yet.

4. **Propolis (Bee glue) :**

The name propolis is derived from two Greek words ; *pro* - before, *polis* - city, because it is often used by the bees in reducing the size of the hive entrance and to seal up and protect their city. Propolis is a gummy, sticky, resinous material collected by the bees from trees and other vegetation, either from buds (like poplar) or from bark (conifers). Bees collect resinous substitute materials like caulking compound or paint when it is needed badly. It consists of different resins collected from buds of horse chest nut, pine, fur, willow etc. Bees collect propolis in their pollen baskets. It sticks to the hands and clothing and is difficult to remove from comb honey and from the frames. But during transportation of bee hives, it helps in holding hive parts together. Bees do not store propolis in the hive but they may redistribute a small quantity time to time.

Uses :

Bees use propolis to fill in cracks and crevices in the hive, to reduce the openings, to strengthen comb attachments etc. When the bees occupy a new hive they coat the inside walls of their home with a very thin layer of propolis. They also coat the inside of every wax cell before use for laying or filling with honey. They use propolis to glue everything together and mix it with wax and fill up the undesirable holes. Sometimes, the bees use propolis of several colours, and an interesting pattern emerges. It is also used to seal the bodies of enemies or other objectionable objects in the walls with bee glue. The dwarf honey bee of South-East Asia *Apis florea* builds the nest at the end of a branch and covers the limb with propolis which serves as a sticky barrier against ants. Thus, the nest is protected.

Propolis has wound healing properties and in the past was widely used as a surgical antiseptic. The folk medicine propolis has antituberculotic properties, it heals wounds and removes corns. The use of propolis ointment prevents the reaction of radiation on skin. The diseases of upper respiratory tract like throat infection, common cold and of the lungs are controlled by inhalation of propolis. The products made up of propolis variously help in mouth and gum disorders, gum decay, general illness, pain from stomach ulcers and fungal skin complaints. In the Soviet Union, propolis is used in veterinary practice in ointments for treating cuts, abscesses and wounds of animals. Russian medical scientists used it for treating burns, external ulcers, eczema in humans and hearing defects.

Propolis was used as the principal constituent in the varnish while making the violins by famous violin makers in Italy.

5. **Bees Wax :**

Bees wax is synthesized by the bees and so can be called a real bee product. That is a natural secretion of the worker bees and is poured out in thin, delicate scales or flakes. It is produced by four pairs of wax glands present on 3rd to 6th abdominal sterna.

Bees wax is very useful by-product of beekeeping industry. It is yellowish to greyish brown in colour and insoluble in water but completely soluble in ether solvent. It is obtained from the combs of wild hives, frame hives and cappings. The main source in India is the combs of the wild bees (*Apis dorsata*) from which several pounds of wax is produced annually. It's melting point ranges from 63.5 to 65°C and specific gravity is 1.

6. Bee Venom :

The worker bees possess a sting at the tip of the abdomen. The sting consists of acid and alkali glands, the secretions of which together form the venom. The 15 day old bee contains about 0.3 mg bee venom. It is a transparent clear liquid, with a sharp, bitter burning taste, an aromatic odour, acidic in reaction when tested with litmus paper and a specific gravity of 1.1313. It dries quickly at room temperature. It is soluble in water and acid. When kept dry its toxicity retains for several years.

Uses :

Bee venom is potentially used in medical practice. It is used in two ways i.e for the treatment of rheumatoid arthritis and for the desensitization of hypersensitive individuals. Its use in treatment of arthritis is centuries old. 'Bee venom therapy' or 'Apitherapy' has long been practiced in Europe and in recent times in United States. Recent investigations show that bee venom acts as an antibiotic and used in different diseases. Bee venom improves blood supply, prevents neuralgic and rheumatic pains and dilates blood vessels. It is an active remedy in cases of acute rheumatic carditis. It also causes a lowering of blood pressure, haemolysis and contraction of muscles. In folk medicine, bee venom has been used to treat certain eye diseases, inflammation of the cornea and mucous membranes. It is also used against skin diseases of face.

3. SILKWORM

Silkworm : The silkworm or silkmoth is an insect which belongs to the order *Lepidoptera* of the class *insecta* and phylum *Arthropoda*. The true silk is obtained from the cocoons of insects popularly known as silkworm. The rearing of silkworms for the production of raw silk is known as *sericulture*.

Silk and silk garments have always been the first choice of the members of the Royal families and other associated with them in India. It is one amongst the oldest business being conducted in our country and a large number of families are associated with the production of silk in India. It is a fact that it has never been a native of India but today the Indian silk is on the top both in quality and quantity.

In India, first of all Lefroy (1904-05) started investigations on silk worm and sericulture at Pusa Institute, New Delhi. Today, it is well known cottage industry achieving continuous progress through improved Technology. India has silk producing centres in Assam, Bengal, Chennai, Mysore, Karnataka and Jammu and Kashmir. India is second in the world for production of Tasar silk.

The modern silk industry in India has grown to meet the domestic rather than export requirements. Further, this industry provides employment opportunities to about 60 lacs people in India. It plays key role in the uplift of rural economy besides earning considerable foreign exchange.

Species of Silkworms : Silk is prepared by the species belonging to family *Bombycidae* and *saturniidae*. *Bombyx mori* is well known silkworm belongs to *Bombycidae* while *Eri, Tasar* and *Muga* moths belong to the family *saturniidae*.

1. **Mulberry silkworm (*Bombyx mori*) :** It is the native of China and now this silkworm has been introduced in all the sillk producing countries like Japan, India, Korea, Italy, France. Its natural food is mulberry leaf so it is called mulberry silk. The colour of silk produced by this silkworm is white or yellow.

2. **Tasar silkworm (*Antheraea paphia*) :** It is commonly known as Japanese oak silkworm and also reared on large scale in Japan. It produces coarse silk. These species feeding on Sal, Asan, Arjun and Ber leaves. These can not be domesticated and are collected from the forest and wild shrubs and trees.

3. **Muga silkworm (*A. assamensis*) :** It is semidomesticated silkworm. The native place of this species is Assam, where now it has become good source of cottage industry. The silk is produced by this species is called muga silk which is of not good quality.

4. **Eri silkworm (*Attacus ricini*) :** Feeding upon the leaves of castor is completely domesticated and produce deep brown coloured silk. It is dull in colour but stronger than cotton threads. The rate of silk production is much slow and quality is of lower grade.

5. **Oak silkworm (*A. pernyi*) :** It is known as Chinese oak silk worm which yields silk in large quantity. The colour of silk is pale buff.

6. **Giant silkworm (*Attacus altas*) :** It is found in India and Malaysia, which is the largest of all the living insects.

4. RED COTTON BUG

Class	–	Insecta
Order	–	Hemiptera
Family	–	Pyrrhocoridae
Genus	–	*Dysdercus*
Species	–	*cingulatus = koenigii* (Fab).

The red cotton bug has wide distribution, it is a minor pest in cotton growing region of northern India particularly Punjab and Uttar Pradesh. This pest also occurs throughout the Maharashtra state but is minor importance. It is commonly known as a "*cotton stainer*".

Host Plants : Cotton, bhendi, ambadi, hollyhock and several other malvaceous plants.

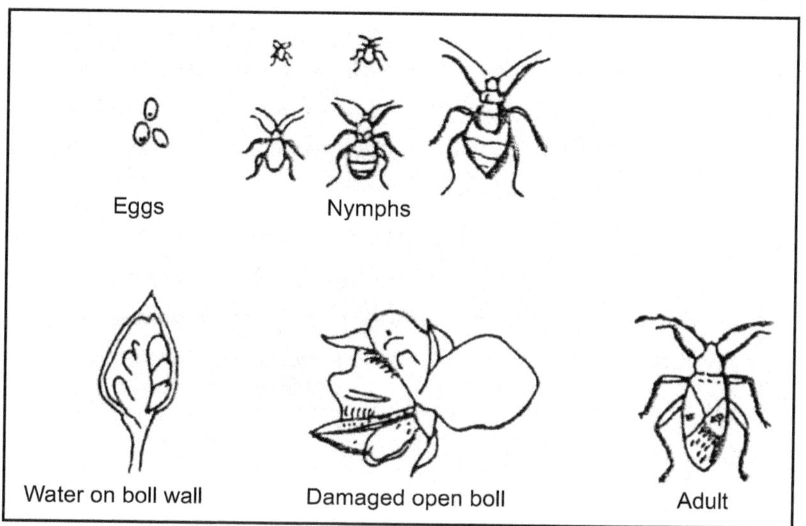

Fig. 3.27: Red Cotton Bug

Economic Importance :

Both nymphs and adults, suck the cell sap from the leaves and tender shoots and impair the vitality of the plant. If the attack is severe, bolls open badly and the lint is of poor quality. In addition they also feed on the seeds and lower their oil content and low percentage of germination; such seeds are unfit for sowing. The lint is stained by the excreta of bugs or by their body juice as they are crushed in the ginning factories.

Control Measures :

1. Cotton field should be ploughed to expose eggs to sunlight.
2. Insects should be hand picked and killed in kerosinised water.
3. The crops of bhendi should be sown as trap crop and pests collected there, should be destroyed.
4. Moistened cotton seeds should be hunged up at different places in the field where bugs congregate, they may get killed in the kerosene mix water.
5. Spraying of Malathion 0.05% is effective to control the pest.
6. Spraying of 1 litre endosulphan 35% EC, 0.25 litre phosphamidon = 100% EC or 1 litre Fenitrothion 100% EC per hectare is very effective or reduces pest population.

5. ANAPHELES MOSQUITO

Systematic Position:

Kingdom:	Animalia	Order:	Diptera
Subkingdom:	Metazoa	Suborder:	Nematocera
Phylum:	Arthropoda	Super family:	Culicidea
Subphylum:	Manibulata	Family:	Culicidae
Class:	Insecta	Genus:	*Anopheles*
Subclass: Pterygota		Species:	*culcifacies*

There are about 30 genera and 2700 species of mosquitoes. The most important mosquito genera are *Anopheles, Culex, Aedes, Mansonia* in India and they are related to disease transmission. Well known Indian species are *Culex pipens, Culex quainque fasciatus (= fatigans), Aedes aegypti, Aedes albopictus, Anopheles culicifacies* and *Anopheles stephensi, Mansonia annulifera* and *Mansonia uniformis*.

Mosquitoes are found all over the world. They only require water, plants and their host species. They occur in deep mines and at high altitudes. They develop in ice cold water and also found in hotsprings. Some of them require only clean water for breeding. Some mosquitoes prefer polluted and filthy water. Many species live in damp forests and marshes. While others are dwellers of crowded human habitations.

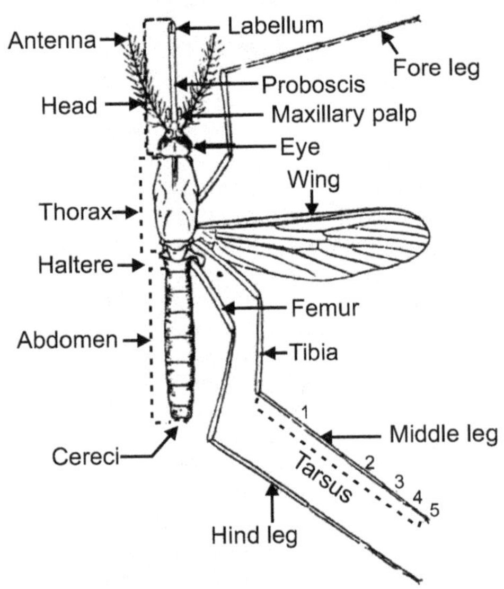

Fig. 3.28: External morphology of a female mosquito

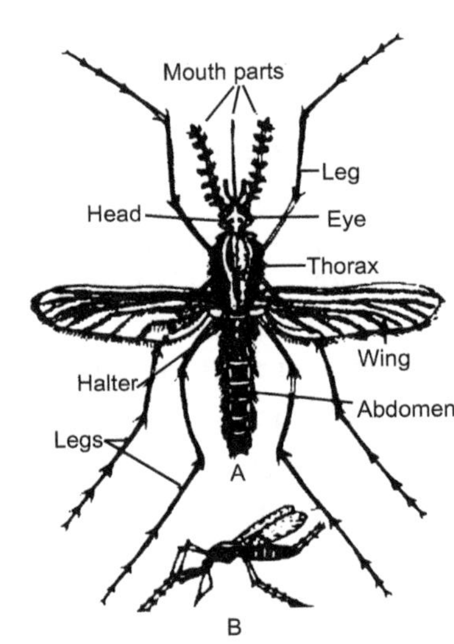

Fig. 3.29: *Anopheles stephens*. Female

A – Dorsal view, B – Sitting posture

Morphology: Body of the mosquito consists of three parts. Head, Thorax and Abdomen. Head bears pair of large compound eyes, long needles like proboscis with which mosquito bites. Pair of palpi (4-joined), situated on either side of the proboscis. Pair of antenna or feelers. They are brushy in male and not quite so in the female. They provide an easy means of distinguishing the male from the female. Thorax bears pairs of wings dorsally and three pairs of legs ventrally.

The abdomen is long and narrow and is composed 10 segments. The last two of which are modified to form the external genitalia.

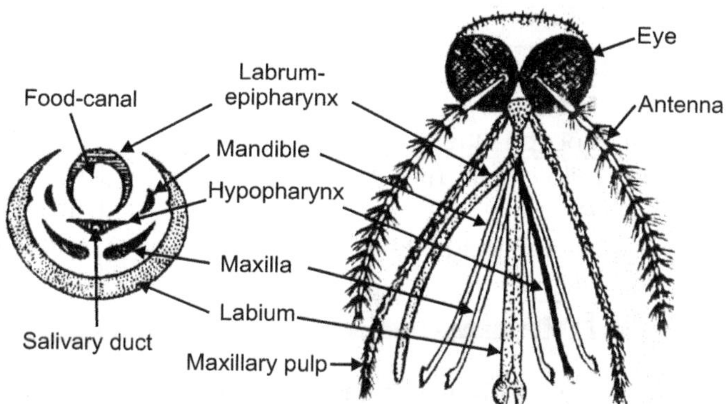

Fig. 3.30: Mouth parts of female *Anopheles* (right) with cross section (left)

Culex	Anopheles	Aedes
1. Body soft, slender and covered with small scales.	Body soft, slender and covered with small scales except the abdominal sterna.	Body soft, slender and covered with small scales.
2. Proboscis not in a line with body.	Proboscis in a straight line with the body.	Proboscis black.
3. Maxillary palps as long as the proboscis in males; shorter and three jointed in females.	Maxillary palps and proboscis of equal size in both the sexes.	Maxillary palps are shorter than proboscis and 3-jointed in female while in males they are equal length and not clubed shaped.
4. Trilobed scutellum with a tuft of bristles on each lobe.	Scutellum crescentic and with bristles along with posterior border.	Scutellum is trilobed and with tuft of bristles on each lobe.
5. Wings with uniform colour and unspotted.	Wings spotted.	Wings unspotted.
6. The body is parallel to the surface at rest.	The body is at 45° angle with the surface at rest.	The body is parallel to the surface at rest.
7. Brownish yellow basal bands present in the abdomen.	–	Abdomen and legs black with white stripes.

Economic Importance of Mosquitoes

Mosquito Borne Diseases:

Mosquitoes play an important role in transmission of human diseases. They act as vectors of many diseases in India.

Types of Mosquito	Disease		Causative agent
1. Anopheles	Malaria		*Plasmodium* species
2. Culex	(a)	Filaria	*Wuchereria bancrofti*
	(b)	Encephalities	Arbovirus
3. Aedes	(a)	Yellow fever (not present in India)	Arbovirus Group B
	(b)	Dengue	Arbovirus Group B
	(c)	Haemorhagic fever	Arbovirus Group B
	(d)	Filaria	*W. bancrofti* (diurenal)
4. Mansonoides	Filaria		*Brugia malayi*

Malaria:

Malaria is transmitted by the bites of certain species of infected female Anapheline mosquitoes. A single female anopheles mosquito during her life time may infect several persons. It is infective when sporozoites are present in the salivary glands otherwise it is not infective.

Following are the conditions necessary for transmission of Malaria:

(1) A reservoir of infection, i.e. presence of individuals with a sufficient number of mature viable, male and female gametocytes in their blood.

(2) Presence of atleast one species of anopheline mosquito which is capable of transmitting malaria.

(3) Climatic conditions should be favourable for the development of sexual cycle in the insect vector and the vector should live long enough to allow the sexual cycle to be completed.

(4) The presence of susceptible human beings to whom the infection may be transmitted.

Apart from mosquito bite malaria may be transmitted by blood transfusion. Malaria may also be transmitted in drug addicts by shared syringes.

The length of time between the bite of an infected mosquito and the first attack of fever is usually not less than 10 days. This is called incubation period. This period varies according to the species of the parasite. Usually about 12 days are required for the development of *P. falciparam* infection, 14 to 15 days for *P. vivax* and *P. ovale* and upto one month for *P. malaria* infection with some strains of *P. vivax* the incubation period may be delayed for as long as 6-9 months.

There are three stages seen in Malaria.

(i) **Cold Stage:** Sudden onset of fever with rigor and sensation of extreme cold accompanied by shivering. The patient desires to cover himself with blankets.

(ii) **Hot Stage:** Patient feels burning hot and casts off his clothes. There is intense headache.

(iii) **Sweating Stage:** Fever comes down with profuse sweating, there is enlargement of spleen and secondary anemia with a tendency to relapses.

Malaria is widespread in the tropical and subtropical regions. The WHO reports there were 198 million cases of malaria worldwide in 2013. This resulted in an estimated 5,84,000 to 8,55,000 deaths the majority (90%) of which occurred in Africa. In India also every year millions of people die due to malaria.

***

Practical **9**...

Earthworm : Vermicomposting Bin Preparation and Maintenance

Aim : To prepare and maintain Vermicomposting Bin (Earthworm)

Kitchen waste is generated daily in large quantities. It is not a waste but a resource at the wrong place at the wrong time. Kitchen waste is mostly organic in nature and can be easily converted to compost by the action of various bacteria, fungi and earthworms, hence the term Bioconversion. Currently, municipal garbage is subjected to bioconversion on a large scale by "*Excel Industries*" at their treatment plant at Chincholi and Deonar. Green cross society has set up a "*Vermiculture project*" at Versova sewage pumping station for the treatment of vegetable waste collected from our Bungalows area. Agricultural college in Pune has popularized the bioconversion of garden leaf litter and farm waste into manure by compositing in a pit. Excel Industries Ltd. has also developed a simple and economical technique to treat kitchen waste and garden leaf litter using mixed bacterial and fungal inoculum. By this process, within 21 days rich manure is ready for use in gardening.

(A) Vermiprocessing of kitchen waste (small scale) at house varanda/garden

(i)   Preparation of Vermiculture Bed : Take a plastic bucket (18-20 litres capacity) and make a drainage hole at the bottom.

- Put 3 inches of soil layer.
- Put 1 inch of the vermiculture layer.
- Again put 1 inch of soil.
- Sprinkle a layer of cowdung and cover it with grass cuttings or leaves of 3 inches thickness.
- Moisten the entire systems gently by sprinkling water, which has been stored in a tub for three days (storage removes chlorine from water).
- As grass and leaves dissolve, add more of the same and continue to maintain 3 inches of this mulch layer.
- Do not allow the system to become soggy due to excess water.
- The system will take 6 to 8 weeks for the earthworms to hatch and stabilise.

(ii) Treatment of kitchen waste using vermiculture Bed

- Spread the kitchen waste on the vermiculture bed.
- First, put waste vegetable greens on the vermiculture bed.
- Rock dust or black sand should be sprinkled every day with the food waste.
- Lime should be sprinkled with fruit peels and non-vegetable waste.
- Bad smell indicates overloading. Insect flies indicate acidity. Allow the vermiculture bed to be airy. If the vermiculture bed is functioning properly, it will not smell.
- Stir the top occasionally. The entire contents can be used as manure within 3 months of treatment.
- Set up a second vermiculture bed using a small portion of the above manure as vermiculture.

Maintenance of the Bin:

1. Feed the worm on one side of the bin for a couple of weeks in order to draw the worms to that side.
2. Once all the worms are on one side harvest the compost on the other side and use it in pots, your garden or sprinkle it across the yard. You can also scoop compost and worms onto a newspaper and sort them out. Be sure to harvest compost at the end of the week before you feed the worms again.
3. If there are too many worms in the worm bin, share extras with friends and family or release some with the dirt in your yard.

Materials Required :

1. Plastic bin.
2. Drill machine.
3. Shredded paper.
4. Soil.
5. Water.
6. Earthworms (different species).
7. Food scraps (vegetables, fruit scraps, bread etc.).
8. Lime or crushed egg shells.

***

Practical **10**...

Visit to Vermicompost Unit

Aim : To Visit a Vermicompost Unit.

Visit the Vermicompost unit of Agriculture University. M.P.K.V. Rahuri (Ahmednagar), Agriculture College, Pune, Yashwantarao Chavan Maharastra Open University, Nashik or any farmer's vermicompost unit of your area; to study the unit with the following guideline points and make report.

(i) Observation of different species of earthworm i.e. *Edrilus eugenae, Eisenia foetida, Perionyx sensibaricus, Perionyx excavatus, Pontoscolex-corethrurus, Polypheretima elongata,* and *Eudichogaster* species.

(ii) Unit : Size of container or pits, shelter, feed mixture, production of vermicompost, collection of worm castings.

(iii) Factors affecting the culturing of earthworms, food, moisture, temperature, light, hydrogen ion concentration, predators and parasites.

(iv) Economic importance of vermicompost (biofertiliser).

***

Practical **1** ...

Study of Microscope : Simple and Compound

Aim : To Study the Microscope : Simple and Compound.

Microscopy :

The word microscope is derived from two Greek words, *'micro'* = small and *'scope'* = to view. Thus, microscope is an instrument designed for visual examination of small object which can not be examined by the naked eye. There are different types of microscopes such as simple, compound, phase – contrast and electron microscope. The study of different microscopes is called *microscopy.*

The earliest recorded use of magnifying lens goes back to Conrad Gesber (1558), a Swiss biologist who published work on the structures of *foraminiferans* (Protozoans). It was Zacharias Janssen, a Dutch Astronomer, who in 1590 added another lens to the telescope and thus provided for the first time a prototype of the present day telescope and the compound microscope.

Anton Van Leeuwenhoek (1674), the father of biology, was the first to use the microscope for biological studies. The compound microscope was constructed by Robert Hooke (1665) and the forerunner of the present day compound microscope.

Simple Microscope (Fig. 4.1)

The simple microscope is commonly used in the laboratory which is nothing but a magnifying glass. It is used when very small animal have to be dissected, or for magnification of the parts in the sections or of small organisms which are not seen by naked eye. The simple microscope is made up of two parts (1) Viewing or optical part and (2) Mechanical part.

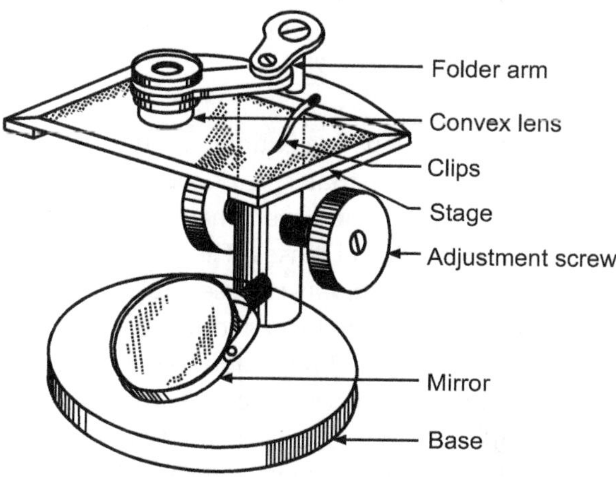

Fig. 4.1: Simple Microscope

Optical part consists of single convex lens. Whereas the mechanical part holds the lens in the proper position. With the help of knob proper magnification can be obtained. Mirror is present for light and clips for holding the slide.

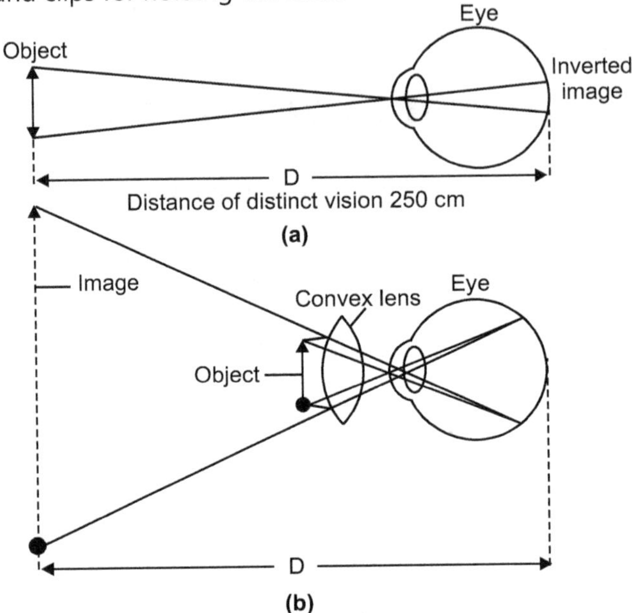

Fig. 4.2: (a) Perception of an Object by the Human eye at a Distance of Distinct Vision. (b) Perception of an Object through a Simple Convex Lens. Which Produce a Magnified Image at the Distance of Distinct Vision.

A simple microscope consists of single lens or a combination of lenses which functions as convex lens. A convex lens magnifies the object and helps to produce a magnified image of a near object which appears to be at the distance of distinct vision. The enlarged image is formed on the retina. It show details which cannot be seen by naked eye. The magnification obtained with a convex lens can be easily calculated by the following formula.

$$M = \frac{250 + 1}{f}$$

Where, f is the focal length of the lens in centimetres and 250 is distance of the distinct vision in centimetres. The magnification that can be obtained with a simple lens cannot exceed three times (3 × magnification). However, magnification more than 3 × 6 can be achieved by using a combination of several lenses. The combination of several lenses fused together acts as a single convex lens. The lenses fused together are called elements. Each element cancels out the distortions and defects produced by the other element. The net result of the combination is the formation of an image free from distortions. A magnification of about 20 times can be obtained. Such microscopes are used by biologists for smaller magnification during field work. It consists of a combination of double concave lens of crown glass fitted between two double convex lenses of flint glass. This is called an achromatic triplet or a planner lens. The three lenses are cemented together and function as one lens.

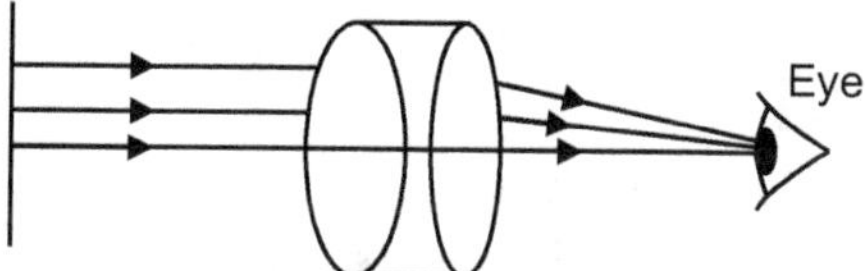

Fig. 4.3: An Achromatic Triple or a Planner Lens Showing the Path of Light Rays while Functioning as a Magnifier

Compound Microscope:

A microscope magnifies the image of an object. The modern compound microscope which is also called light microscope, is most important optical instrument for the study of cells and tissues. It is important apparatus in a medical laboratory. It is a precision instrument and needs careful handling. Improper use of the microscope leads to the loss of clarity of the image.

The light microscope uses white light, either the external sunlight or the internal tungsten filament lamp, as a source of illumination. While observing under microscope, the objects either look dark or coloured, contrasted against a lighted background.

The Microscope and its Accessories:

Accessories or components of commonly used light monocular microscope is shown in Fig. 4.4. All the components of monocular and binocular microscopes are the same except that the binocular microscope has two eye pieces (ocular), which allow the user to keep both eyes while viewing through the microscope. Both the microscope have three main systems such as support system, illumination system and magnification system.

I. Support System: It consists of the following parts:

　(i) **Foot Rest:** It is used to hold the microscope while transporting.

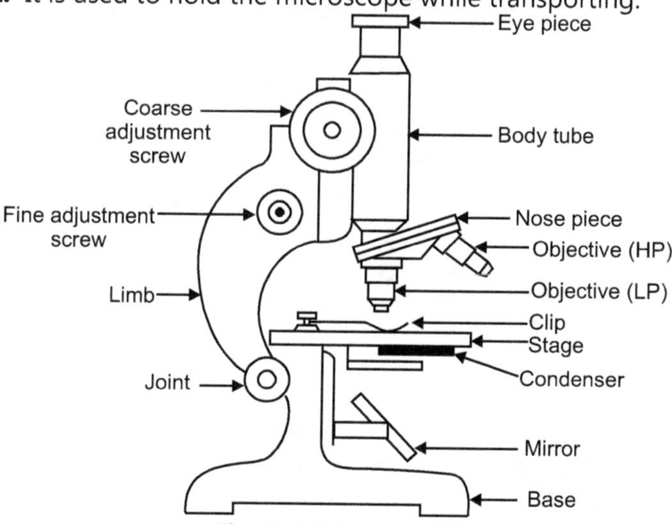

Fig. 4.4: Microscope

　(ii) **Tube:** It holds the optical system for magnification. It may be straight and fixed (older models), or bent and movable (modern models) for convenient viewing. The tube has eye piece at upper end i.e. near the eye. In case of bent tube, a special prism is put in the middle of the tube where it is bent. The prism directs the light beam coming through the objective lenses towards the eye piece. In binocular microscope, the light beam further shifts and reaches both the eye pieces.

(iii) Stage: It is used to hold the object slide in place for viewing. The stage has central hole for passing the light in order to illuminate the object. The stage can be fixed or a mechanical stage can be screwed on to the fixed stage which can move the object slide across the stage or along the stage, horizontally or vertically. Focusing of optical system can be done either by moving the stage up and down (older models) or by the movement of the limb (modern models). The support system helps to carry the microscope in an upright position with one hand holding the base.

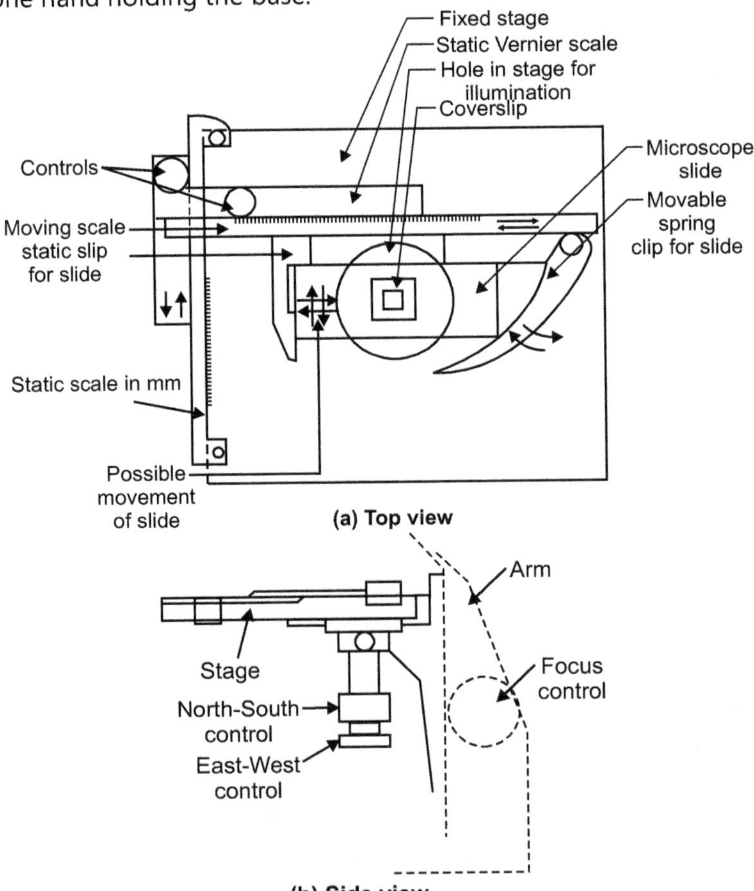

Fig. 4.5: Parts of the Mechanical Stage from (a) Top View, and (b) Side View

II. Illumination System:

This system provides proper illumination to the object. It consists of the following parts:

(i) The Source of Light: Electric light should be preferably be used, since it is easier to adjust. It is provided by the lamp, placed in front of the microscope. An opaque electric bulb of the daylight type of 60 watts is used. Otherwise, daylight can be used.

(ii) Mirror: The mirror reflect rays from light source on the object. One side has a plane surface and the other side has a concave surface. The concave side forms a low power condenser and is used if there is no condenser.

(iii) The Condenser: The condenser brings the rays of light to a common focus on the object to examine. It is situated between mirror and the stage. It can be raised (maximum illumination) and lowered (minimum illumination). The common type of condenser consists of two or three lenses.

(iv) The Diaphragm: The diaphragm which is lodged below the condenser is used to reduce or to increase the angle and therefore, also the amount of light that passes into the condenser. The upper lens of the condenser coincides in position with the aperture on the stage.

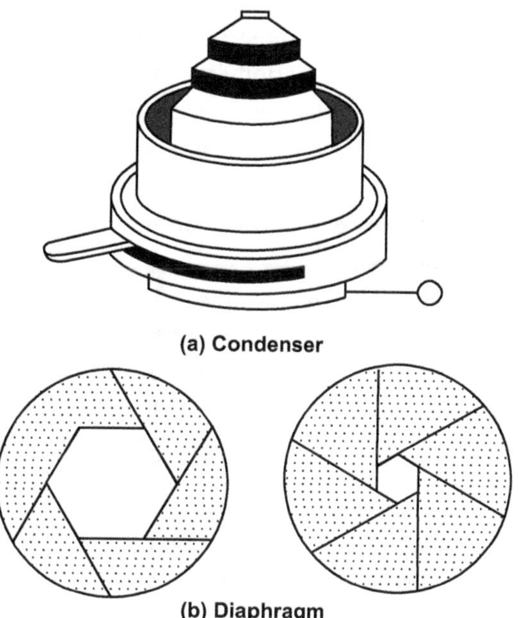

Fig. 4.6: Parts of Illumination System

III. The Magnification System:

This consists of a system of lenses. The lenses of the microscope are mounted in two groups, one at each end of the long tube - the body tube. The first group of lenses is at the bottom of the tube, just above the preparation under examination (the object) and is called the *objective*. The second group of lenses is at the top of the tube where the microscope applies his eye and is called the eyepiece.

The Objective: The magnifying power of each objective is shown by a figure engraved on the sleeve of the lens.

×10 objective magnifies	10 times (low power)
×40 objective magnifies	40 times (high power)
×100 objective magnifies	100 times (oil immersion)

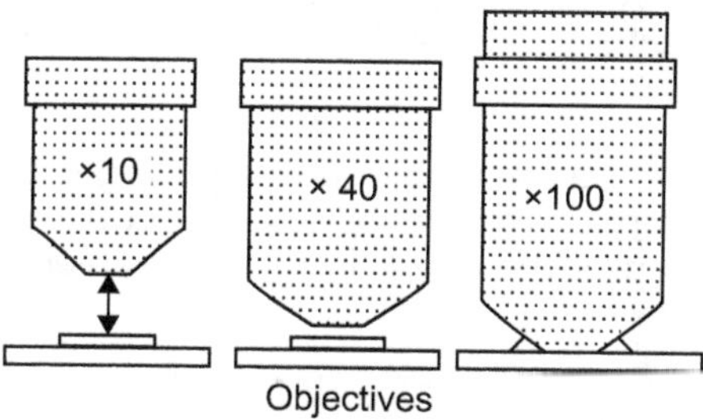

Objectives

Fig. 4.7: Objectives of the Compound Microscope

The ×100 objective is usually marked with a red or black ring to show that it must be used with immersion oil.

The greater magnifying power of the objective, the smaller working distance i.e. distance between the front lens of the objective and the object slide when the image is in focus.

The eyepiece : The magnifying power of the eyepiece is marked on it.

- ×5 eyepiece magnifies the image 5 times
- ×10 eyepiece magnifies the image 10 times

If you see an ×40 objective with an ×5 eyepiece, the object will be magnified 40 × 5 times, i.e. the total magnification will be 200.

IV. The Adjustment System:

The system comprises of the following:

1. **Coarse Adjustment Screw:** This is the largest screw. It is used first to achieve an approximate focus.

2. **The Fine Adjustment Screw:** This screw moves the objective more slowly. It is used to bring the object into perfect focus.

3. **The Condenser Adjustment Screw:** This is used to raise the condenser for greater illumination or to lower it to reduce the illumination.

4. **The Iris Diaphragm Lever:** This is small lever, fixed on the condenser. It can be moved to close or open the diaphragm, thus reducing or increasing both the angle and the intensity of the light.

5. **Mechanical Stage Controls:** These are used to move the object slide on the stage. One screw to move it backward and forward; and one screw to move it to left or right.

How to use Microscope ?

Following points should be remembered while using the microscope:

1. Place the microscope on a firm, level bench of adequate size but not too high.
2. Screw the objective into the revolving nose – piece following this order in a clockwise direction.
   × 10 objective.
   × 40 objective.
   × 100 objective.

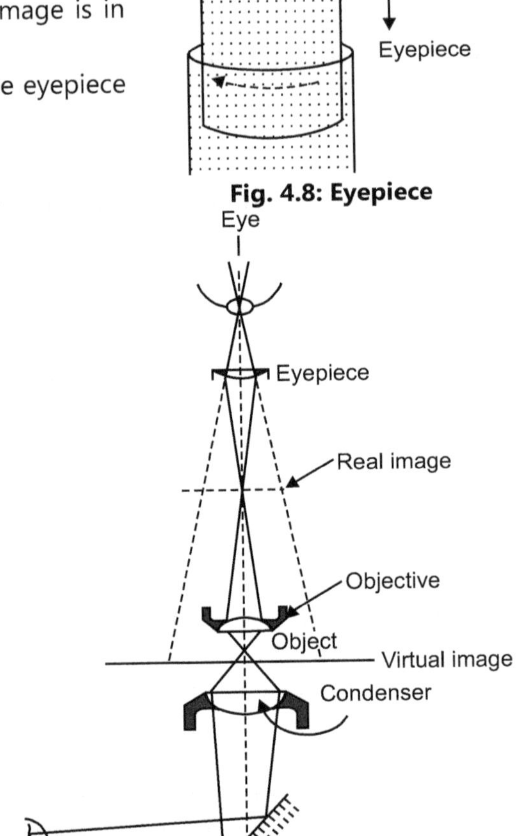

Fig. 4.9: The Path of the Light Waves to Form the Real Image Formed by the Objective. The Eye Sees the Magnified Virtual Image Formed

3. Place the eyepiece in the body tube.
4. Fix the condenser under the stage.
5. Fix the mirror on the foot.
6. If electric illumination is to be used, place the lamp 20 cm in front of the microscope facing the mirror, which should to be fixed at an angle of 45°. Adjust the position of lamp so that it shines on the center of the mirror.
7. Use the plane side of the mirror. Open the iris diaphragm to the maximum. Raise the condenser. Place piece of thin white paper over the lens at the top of the condenser. This piece of paper should show an image of the electric bulb, surrounded by a circle of light. Adjust the mirror so that the image of the bulb is in the exact center of the circle of light (or the brightest part if daylight is being used).
8. It is very important to entire the condenser correctly.
9. Use the low power objective.
   Lower the condenser down to the bottom. Lower objective until it is just above the side preparation. Raise the objective, using the coarse adjustment screw, until a clear image is seen in the eyepiece. Occasionally a clear image can not be obtained although the objective has been lowered as far as possible. This is because the fine adjustment screw has been termed right to the end. Turn it back as far as it will go in the other direction and then focus by raising the objective.
10. Using high power objective (×40): Lower the condenser half way down. Lower the objective until it is just above the slide preparation. Using coarse adjustment, raise the objective very slowly until blurred image appears on the field. Bring into focus, using the fine adjustment. Raise the condenser to obtained sufficient illumination.
11. Using the oil immersion objective (×100): Perfectly dry stained preparation must be used. Place a tiny drop of immersion oil on the part to be examined (use synthetic oil, which do not dry, in preference to cedar – wood oil, which dries quickly).

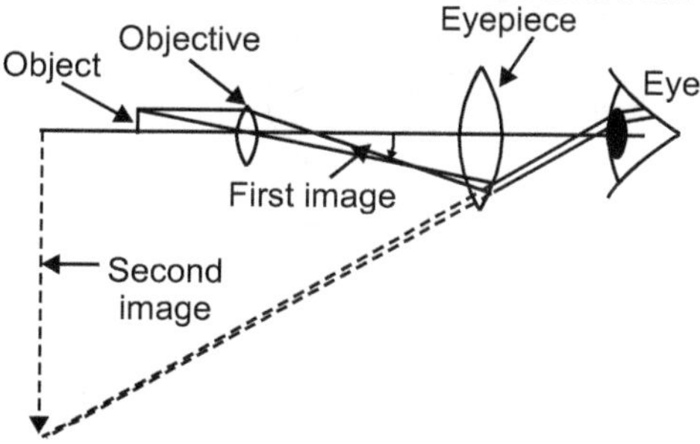

Fig. 4.10: Image Formation in a Light Microscope Occurs in Two Stages

Raise the condenser up as far as it will go and open the iris diaphragm fully. Lower the ×100 objective until it is in contact with the oil. Bring it as close as possible to the slide, but avoid pressing on the preparation. Look through the eye piece and turn the fine adjustment very slowly upwards until the image is in focus. If the illumination is inadequate, use the concave side of the mirror.

12. Remember that the circle of light seen in the eyepiece is called microscope field.
13. Objective seen at the bottom of the field are actually at the top. Objective seen on the left hand side of the field are actually on the right. If you move the slide to the right, the object examined moves to the left. If you move the slide towards you, the object examined moves away from you.

***

Practical **2**...

Micrometry : Measurements of Microscopic Objects

Aim : To Study about Micrometry : Measurements of Microscopic Objects.

Micrometry :

1. Study of Units like Micrometer, Nanometer and Angstrom:

Microscopic objects like chromosomes, cells etc. are measured with the help of the compound microscope.

For the accurate measurement of any microscopic object the stage micrometer and ocular micrometer are used in combination. This field of science is called micrometry. The measurement are expressed in micrometers (μm) and are made with the help of an ocular or eyepiece micrometer disc and a stage or object micrometer.

The eye-piece or ocular micrometer is a transparent circular glass disc which is placed in the body of the ocular. The upper surface of the eyepiece micrometer is either engraved with a scale or divided into a number of equal squares by cross – lines ruled upon it. It is placed in between lenses of the eye-piece in such position that the scale can be distinctly on looking through the microscope.

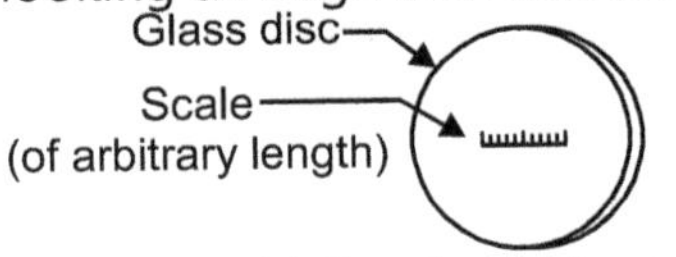

(a) Eyepiece micrometer

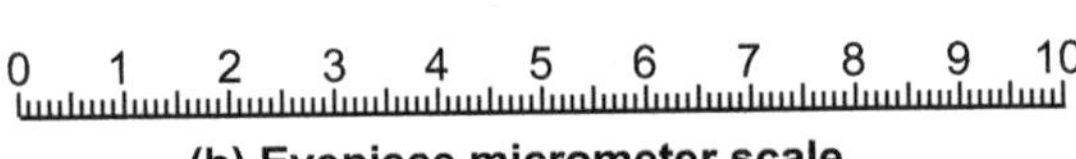

(b) Eyepiece micrometer scale

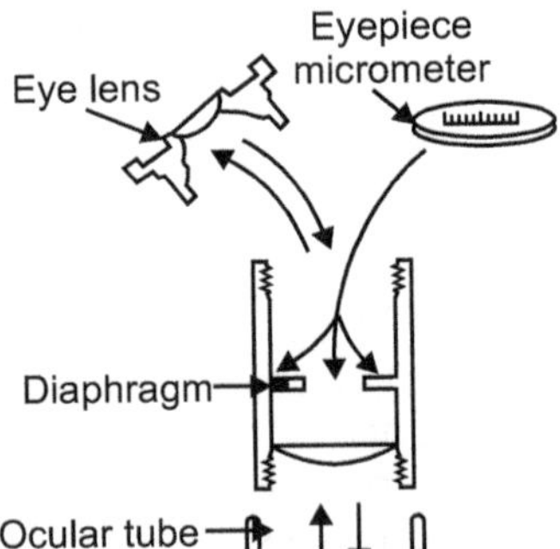

(c) A Longitudinal Section of Eye-Piece to show the Position of Eye-piece Micrometer in it

Fig. 4.11

Stage micrometer is a glass slide on which a scale of known intervals is marked. When the scale of the object or stage micrometer is focused; the ocular micrometer is also seen in the superimposed condition. When the two are graduated in the same unit, the ocular micrometer has to be calibrated against the object micrometer. Calibration is done for each objective of the microscope separately, beginning at the lowest magnification and focusing the scale of the state micrometer. Most stage micrometers have a 2 mm long scale divided into 0.01 mm units. The mechanical stage is aligned with the zero reference line of the stage micrometer so that it coincides with the zero line of the ocular micrometer as shown in the Fig. 4.12.

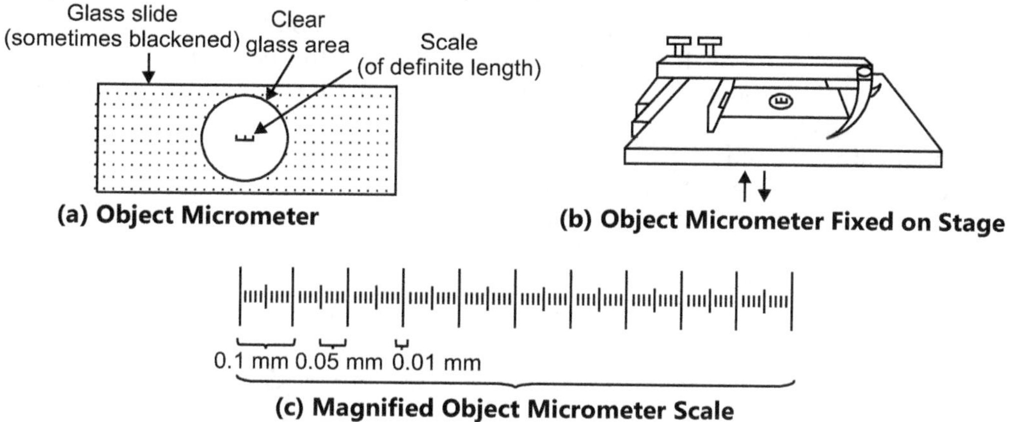

Fig. 4.12

The number of millimeters that correspond to a full scale on the ocular micrometer is first recorded. The distance is then divided by the number of divisions in the ocular micrometer scale and converted to micrometers. In the microscopic field if the full scale on the ocular micrometer coincides with 0.685 mm on the stage micrometer and as there are 100 divisions on the ocular micrometer scale, each division equal 0.685 mm divided by 100 or 6.85 µm per ocular micrometer division. This calibration is repeated for each objective magnifcation of the microscope is noted. Whenever, the size of an object is to be a measured the size of the object is set against the ocular micrometer scale in the eyepiece. The reading on the ocular micrometer is noted and converted into the actual size of the object in micrometers (µm).

Unit	Symbol	Value
1 centimeter	cm	1/10 mm or / 100 metre = 10^{-2} metre
1 millimeter	mm	1000 µ or 1/10 cm = 10^{-3} metre
1 micro or micrometer	µ or µm	1000 mµ or 1/1000 mm = 10^{-6} metre
1 millimicron Or 1 nanometer	nm	$10A°$ or 1/1000 µ = 1×10^{-9} metre
1 Angstrom	A°	0.1 mµ or 0.1 mm or 1×10^{-10} metre

Camera Lucida :

The camera lucida is an accessory instrument of the compound microscope and is attached to the top of the monocular body tube of the microscope. It is used to draw clear simple and exactly proportionate outline sketches of the objects under study. In the biological work sketches are important and hence this device is commonly used, which was invented in 1807 by W.H. Wollaston.

The camera lucida consists of a mirror and reflecting prism which is mounted over the ocular (eye - piece). The surface of the prism is coated with silver plating except in a particular area in the center which coincides in position with the monocular eye - piece. It has an adjustable mirror inclined at 45° both to the desk top and the prism. Light from the paper placed on desk top is reflected by the mirror on to the reflecting face of the prism, which reflects it into the eye of the observer. The light rays passing through the ocular are deflected at 90° by the prism. The rays are further deflected at 90° by the mirror. This arrangement projects an imaginary image on the desk top beside the microscope. If a piece of paper is kept there, the observer can see the superimposed pictures of the paper and the image of the object through the ocular. Observer can see the pencil point. This helps to trace the outline of the image falling on the paper easily by the observer while looking through the eye - piece.

Sketching with the help of camera lucida is not so easy. Certain precautions should be taken, such as avoid bright light passing through object. The bright light makes the pencil point invisible. Even if the paper is bright the object disappears altogether from view. To overcome this problem, many camera lucida are provided with devices like neutral glass filters which can cut the light from either the object or the paper or both. It is advised to use variable resistance in series to the microscope lamp. This helps to reduce the intensity of light. While drawing the figure it is better to use fairly hard pencil which can give precision of line. For record purpose a free hand sketch is sufficient, but for publication in research or scientific journal very accurate and in proportion sketches are required.

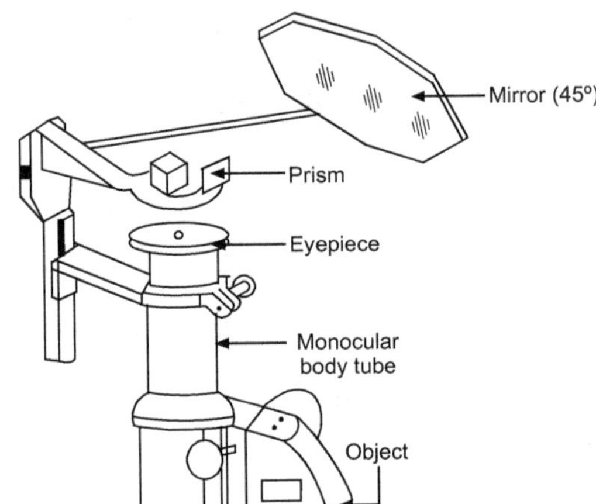

Fig. 4.13: Camera Lucida, Mirror Type

Hence, magnification of the drawing should also be known. For this purpose use of micrometer eye piece to measure the object in various directions is essential. Part of the millimeter scale being drawn with the same set up is used for the drawing. While doing this the angle of inclination of the mirror be should not be altered. If the angle of the mirror is changed the magnification of micrometer scale will change and it will not match with the drawings.

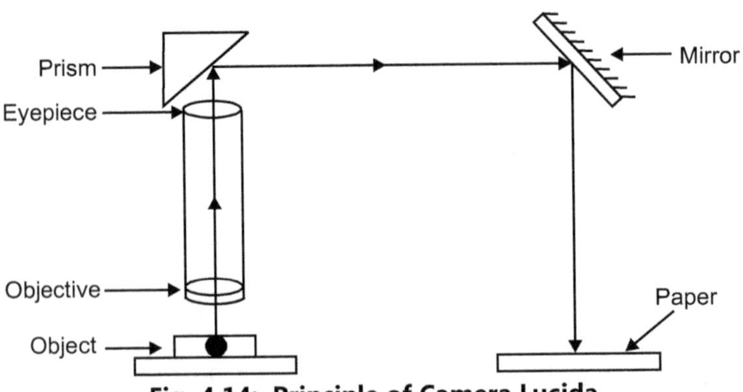

Fig. 4.14: Principle of Camera Lucida
****

Practical 3...

Study of Cell : Preparation of Temporary Mount of Human Buccal Epithelial Cells

Aim : To Study the Cell : Preparation of temporary mount of human buccal epithelial cells.

Materials : Glass microscope slides, coverslips, paper towels or tissue. Methylene blue solution (0.5% to 1%) (mix approximately 1 part stock solution with four parts of water).

Plastic pipette or dropper, sterile, individually packed cotton swabs.

Methods :

1. Take a clean cotton swab and gently scrap the inside of your mouth.

2. Smear the cotton swab on the centre of the microscope slide for 2-3 seconds.

3. Add a drop of methylene blue solution and place a coverslip on top. Note that concentrated methylene blue solution is toxic if ingested.

4. Remove any excess solution by allowing a paper towel or bottling paper to touch one side of the coverslip.

5. Place the slide on the microscope with 4x or 10x objective in position and find the cell. Then view at higher magnification.

   Methylene blue stains negatively charged molecules in the cell including DNA and RNA. The dye is toxic when ingested and it causes irritation when in contact with the skin and eyes.

Observation :

The cells seen squamous epithelial cells from the outer epithelial layer of the mouth or check. The nucleus contains DNA and RNA hence it take dark blue colour and the small blue dots are bacteria from our teeth and mouth.

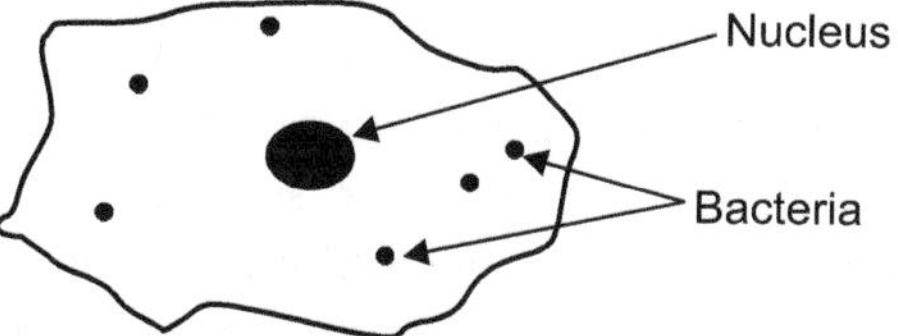

Fig. 4.15 : Squamous epithelial cell

****

Practical **4**...

Preparation of Blood Smears to Observe the Blood Cells

Aim : To Study about Differential Leucocyte (W.B.C.) Count.

Introduction :

Blood : It is an essential fluid which carries our critical functions of transporting oxygen and nutrients to our cells and getting rid of carbon dioxide, ammonia and other waste products. Blood is highly specialised tissue composed of more than 4,000 different kinds of components. Four of the most important ones are red blood cells (RBCs), white blood cells (WBCs), platelets and plasma.

RBCs or erythrocytes are relatively large, microscopic cells without nuclei. They form 40-58% of total blood volume. They are produced in our bone marrow from stem cells at a rate of about 2-3 million cells per second. Haemoglobin is a gas transporting protein. In anemic people, there is deficiency of blood cells. The red colour of the blood is primarily due to oxygenated red cells. While studying the blood smear one can see the RBCs, and WBCs easily with strain.

At birth total RBCs count varies from 6.5 millions to 7.25 millions/Cu.mm.

Leucocytes : These are nucleated cells. In healthy adults, the number of leucocytes is between 5,000 to 10,000/cu.mm (µl) of blood. The various types of WBCs found in circulating blood are given below.

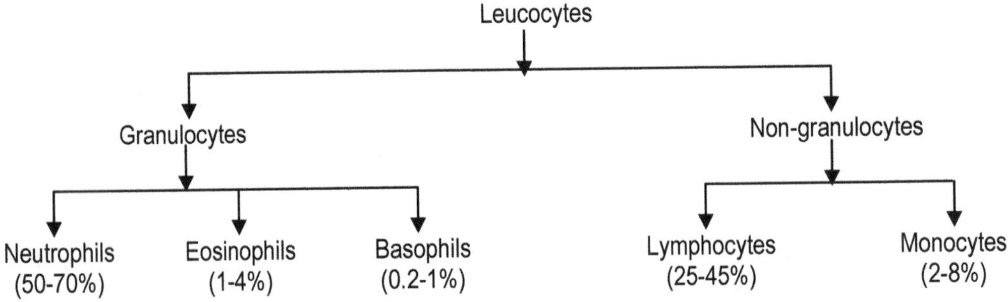

Functions :

Leucocytes are useful for the defense against bacteria, viruses, parasites and cancer cells. The neutrophils and monocytes that attack and destroy viruses and bacteria in the circulating blood. Monocytes are called macrophages which are capable of combating disease agents. Plasma cells are derived from lymphocytes, produce immunoglobulins which neutralize the toxins produced by invading organisms. Lymphocytes are useful for immunity. The eosinophils rise in allergic disorders and parasitic infections. These cells ingest antigen-antibody complexes, neutralize histamine and dissolve clots. The basophils of blood contain histamine, serotonin and heparin. These are powerful vasodilators.

Significance :

Differential leucocyte count is useful to study the changes in the distribution of white blood cells which may be related to the specific type of disorders. It also gives idea regarding the severity of the disease and the degree of the response of the body.

1. **Neutrophilia :** Increase in the percentage of neutrophils called neutrophilia and decrease in neutrophils is called neutropaenia. Physiological causes are responsible for neutrophilia and pathological causes like bacterial, viral infections cause decrease in neutrophils. The other causes like anemias and bone marrow suppression due to drugs and radiation.

2. **Lymphocytosis :** It may be relative or absolute. In case of relative the number of lymphocytes is unchanged but due to decrease in neutrophils differential count shows an increase in lymphocytes. Absolute lymphocytosis observed in children due to certain infections like mumps, cough, measles and certain chronic diseases.

3. **Eosinophilia :** It is increase in eosinophils and observed in asthama, hypersensitivity reactions, parasitic infections and chronic inflammatory diseases.

4. **Monocytosis :** It is observed in malaria, tuberculosis, typhoid and kala azar.

5. **Basophilia :** It is normally observed in chronic myeloid leukaemia.

Normal values (Male or female) :

Neutrophils	40-75% (Mean 57%)
Lymphocytes	20-45% (Mean 37%)
Eosinophils	1-4% (Mean 2%)
Monocytes	2-8% (Mean 6%)
Basophils	0-1% (1%)

Requirements :

Microscope slides, cedar wood oil and reagents.

Principle :

The polychromic staining solutions (Leishman, Giemsa, Wright) contain methylene blue and eosin. These basic and acidic dyes induce multiple colours when applied to cells. Methanol acts as a fixative and also as a solvent. Due to fixative cells adher to the glass slide. The cytoplasm of the WBCs is stained by acidic dye and hence they are called acidophilic or eosinophilic cells. The acidic component i.e. nucleus and nucleic acid takes blue to purple shades by basic dye and hence they are called basophilic cells. The neutral components of the cell are stained by both the dyes.

Procedure :

Prepare a thin smear by spreading a small drop of blood evenly on a slide. This can be done by putting a drop of blood on clean and dry slide at the edge. Place a spreader slide at an angle of 30° to 35°. Pull back the spreader until it touches the drop of blood. Let the blood run along the edge of the spreader. Push the spreader forward to the end of the slide with a smooth movement. Dry the blood smear at room temperature. Then stain the slide. Methanol in stain fixes the smear. For staining cover the smear with staining solution by adding 10-15 drops on the smear. Wait for one minute. Add equal number of drops of buffer solution and mix the reaction mixture adequately by blowing on it through a pipette. Wait for 10 minutes. Wash the smear by tap water. Stand the slide in a draining rack or on the laboratory counter to dry.

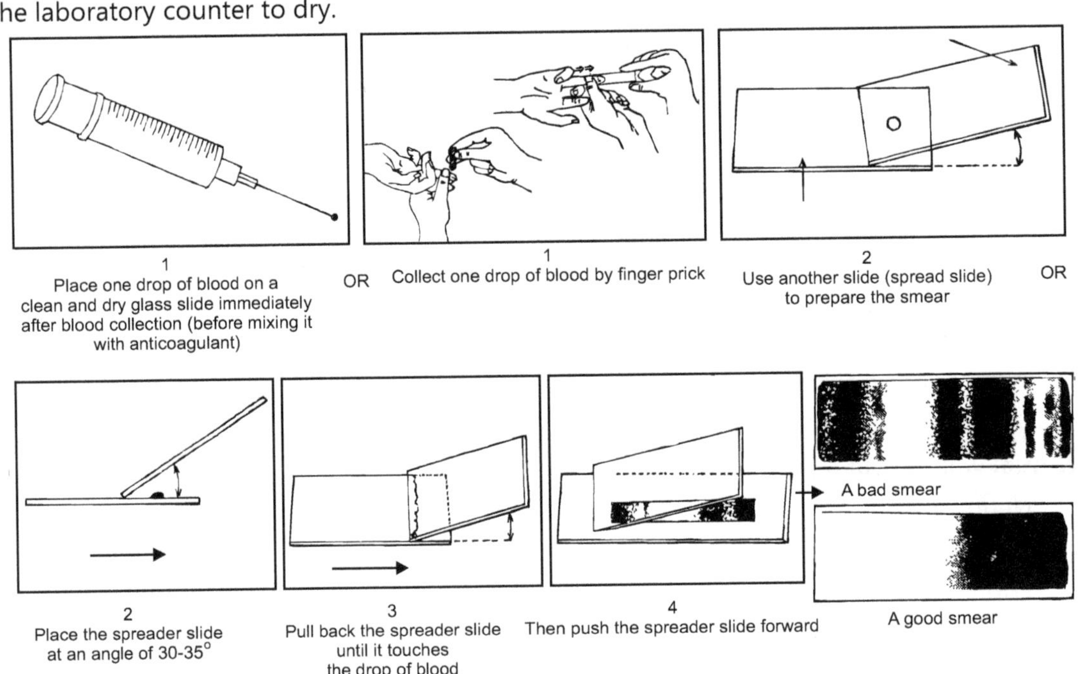

Fig. 4.16

Observation of Film :

First examine the smear under the low power. Select the thin area of smear. Place a drop of immersion oil on the smear. See under oil immersion objective. Examine the film by moving from one field to the next systematically. Record the white blood cells (Leucocytes) seen each field. Count atleast a total of 100 leucocytes. Counting 500 leucocytes gives high degree of accuracy.

Preparation of Stains :

1. **Leishman's Stain :** It is prepared by 0.15 g of powdered stain in 100 ml of methyl alcohol.

2. **Geimsa's Stain :** Take 0.3 g of Geimsa staining powder and dissolve in glycerol and keep at 56-60°C in water bath for 2 hours. Then add 25 ml of acetone free methyl alcohol. Keep the solution at room temperature for one week. Filter it and store in amber coloured bottle at room temperature.

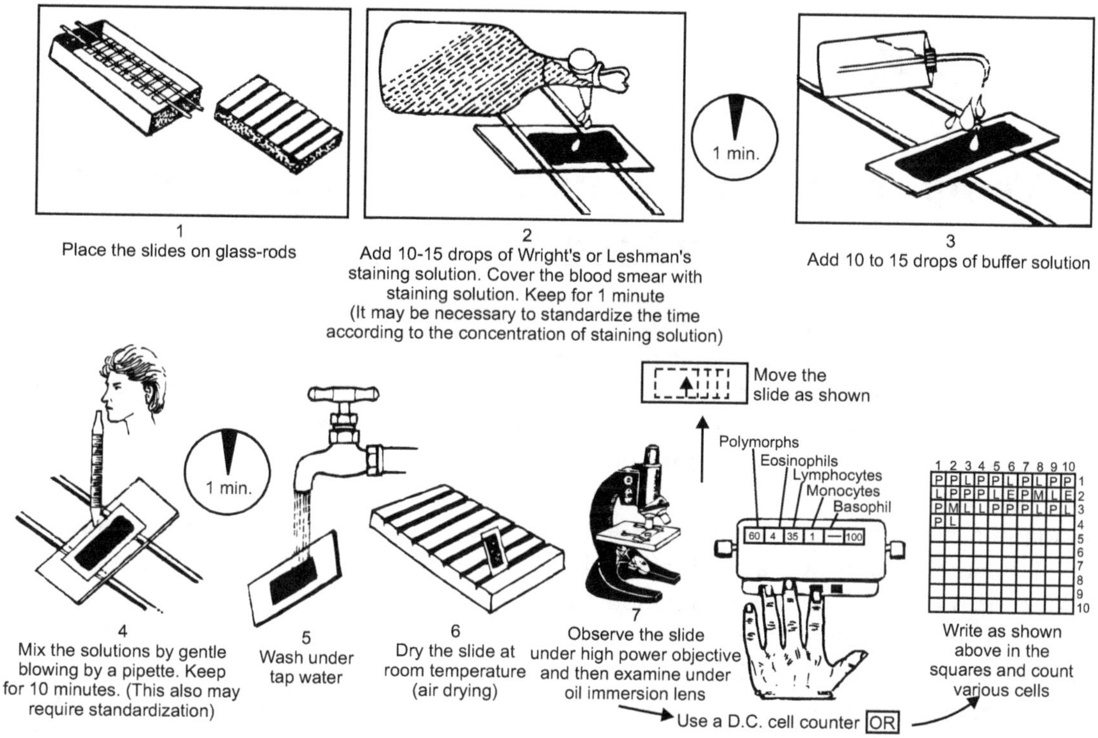

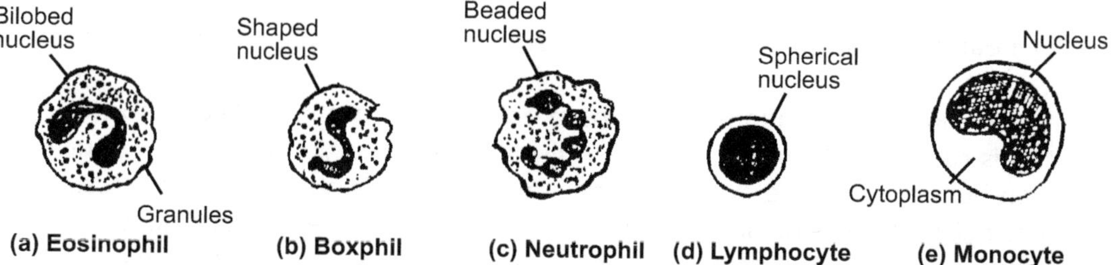

Fig. 4.17

Bilobed nucleus

Shaped nucleus

Beaded nucleus

Spherical nucleus

Nucleus

Granules

Cytoplasm

(a) Eosinophil   (b) Boxphil   (c) Neutrophil   (d) Lymphocyte   (e) Monocyte

Fig. 4.18 : Different types of leucocytes

***

Practical **5**...

Temporary Preparation of Mitotic Cell from Onion Root Tip Cells

Aim : To prepare Mitotic Cell From Onion Root Tip Cells.

Requirements : Carmine, orcein, glacial acetic acid, ethyl alcohol, reflux condenser, conical flask, distilled water, old dried onion bulb, garlic clove, glass jars, petri dish, spirit lamps, blade, needles, fine forceps, slides, coverslips, blotting paper, pins, scissors etc.

Acetocarmine : Used for chromosome staining.

Preparation : Reflux 500 mg carmine with 45 ml of glacial acetic acid. If a reflux condenser is not available, simmer the mixture in a conical flask in fume cupboard. After 15-30 minutes cool it and add 55 ml of distilled water. Filter the solution before use. The acetocarmine stain stored for long period gives better results.

Aceto-orcein : Used for chromosome staining. Aceto-orcein stain can be used in place of acetocarmine. It also gives better results.

Preparation : Add 2 gm of orcein to 45 ml of glacial acetic acid. Further follow the method used for acetocarmine preparation. The stain has a tendency to precipitate and therefore filter it before use.

Acetic alcohol (Fixative) : Acetic alcohol precipitates chromatin, therefore, it is used to fix tissues before chromosome staining.

Preparation : One part by volume glacial acetic acid to three parts of ethyl alcohol (1 : 3).

Stained squash preparation of root tips of onion/garlic for the study of mitosis.

Mitosis can be well studied in actively dividing cells. The tip of root shows such zone. The onion bulb or the garlic clove is an ideal material for demonstration of mitosis. In this plant material, roots develop actively in short period under laboratory conditions.

Procedure : Remove outer peels from onion bulb and with the help of blade carefully cut the dried root from the selected bulbs. To set the material, keep the bulbs on glass jar containing tap water.

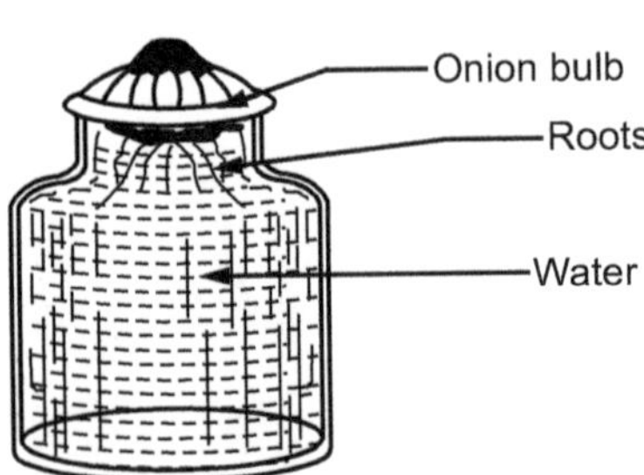

Fig. 4.19 : Experimental set up for mitosis (onion bulb)

When garlic bulb is used in place of onion, the cloves are to be separated and then peeled carefully. Then a pin is to be inserted across the middle part of the clove and is made to stand along the edge of a petri dish containing tap water. Within 24 hours roots appear from onion bulbs or garlic cloves. This material is used for squash preparation. About ten such sets are necessary for a batch of 30 students.

Following are the important steps to obtain better squash preparation from root tips.

(i) Excise roots one centimetre away from the root tip and transfer them to freshly prepared acetic alcohol (fixative). Material should be fixed at least for one hour or preferably for longer period.

(ii) Treat these root tips with 1N HCl at 60°C for a period of 5 minutes.

(iii) Transfer root tips to a cavity block containing acetocarmine or aceto-orcein. Keep the material for staining for about 10 to 30 minutes.

(iv) Transfer the root tip on a clean slide and put a drop of stain on it. Now retain only $1/2^{th}$ part of root tip intact and remove the rest part. Place a clean coverslip over the material and hold the slide in a fold of blotting paper.

(v) Press the coverslip gently with a tip of a needle. This will initiate separation of the cells from the root tip.

(vi) Then remove the fold of blotting paper carefully from slides. To preserve the squash preparation for longer period, it is necessary to seal the edges of coverslip with molten paraffin or nail polish. It will prevent drying of the material.

Scan the preparation under compound microscope. First view it under low magnification and then adjust it under high magnification to locate stages of mitosis in your preparation.

Observations : In the squash preparation, cells in interphase and different stages of mitosis are to be observed. The structure of chromosomes and their orientation in dividing cells, spindle fibres and such other microscopic observations of the cell inclusions help in identifying different stages of mitosis.

Interphase :

(i) Nucleus appears prominent.

(ii) Chromosomes appear as long, thin thread like chromatin fibres.

Prophase :

(i) Chromosomes appear as thin coiled threads. Each chromosome is made up of two chromatids.

(ii) In late prophase, chromosomes appear much condensed and thick bodies. Chromosomes are arranged at the periphery of the nucleus.

(iii) Nuclear membrane and nucleolus disappear on the termination of prophase.

Metaphase :

(i) The thick and short chromosomes appear to be arranged along the equator and are seen attached to the spindle fibres.

(ii) Spindle fibres make their appearance between two poles to form spindle apparatus.

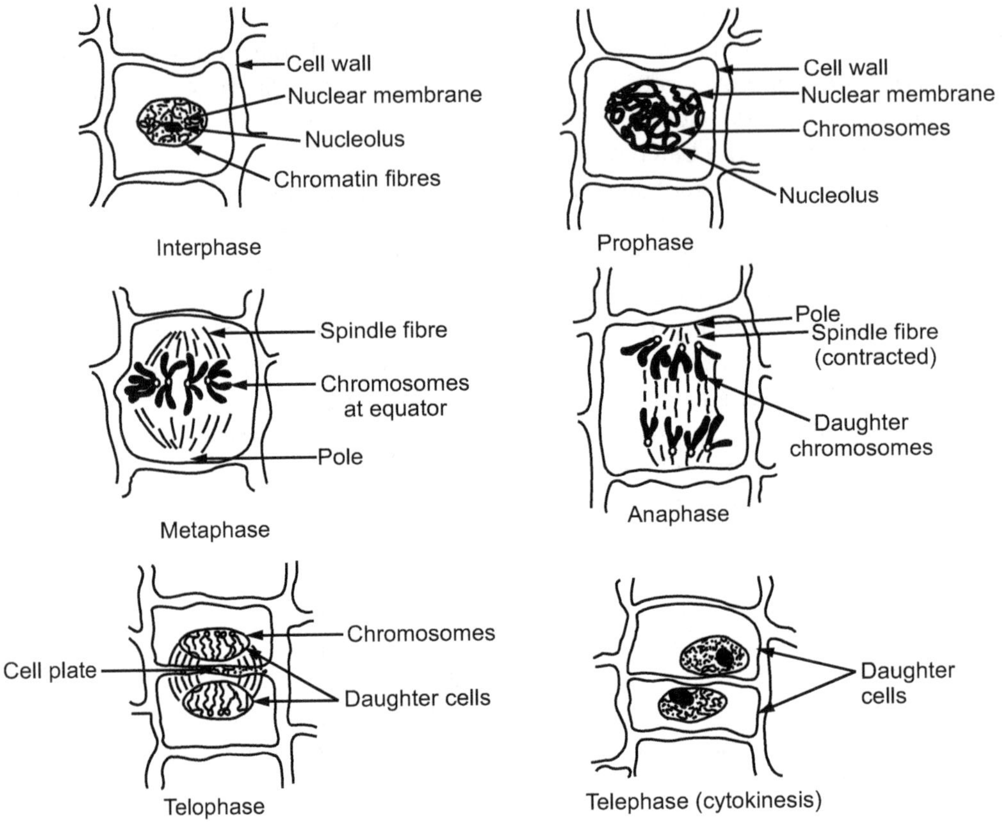

Fig. 4.20 : Stages of Mitosis

Anaphase :

(i)   The sister chromatids of the chromosome get separated longitudinally by splitting of centromere and move towards the opposite poles of the cell due to contraction of spindle fibres.

(ii) Each chromosome appears as inverted 'v' or like 'L'.

Telophase :

(i)   The cells show two daughter nuclei, one nucleus is observed at each pole of the cell. Spindle apparatus disappears.

(ii) A cell plate is seen in the late telophase.

✳✳✳

Practical **6**...

Study of Cell Organelles from Electron Micrographs

Aim : To Study any three cell organelles from electron micrographs.

1. *Endoplasmic Reticulum (E.R.)*

Comments :

1. Endoplasmic reticulum is found in all eukaryotic cells except R.B.Cs. and mature egg cells.
2. E.R. in secretory cells like pancreas, liver and endocrine glands are highly developed.
3. It is a network of tubules, cisterane and vesicles traversing in the cytoplasm; cisternal membranes are interconnected by interconnecting tubules.
4. E.R. is either smooth or rough; the tubular surfaces of smooth endoplasmic reticulum are devoid of ribosomes, while those with ribosomes are rough or granular endoplasmic reticulum.
5. It was first reported by Porter in 1948.
6. E.R. provides mechanical support to the cell, provides increased surface area for cell metabolism and attachment of ribosomes which are the sites of protein synthesis.
7. ER also gives rise to the other membranous structures of the cell like mitochondria and nuclear membrane, and helps in conducting intracellular informations.

2. *Mitochondria*

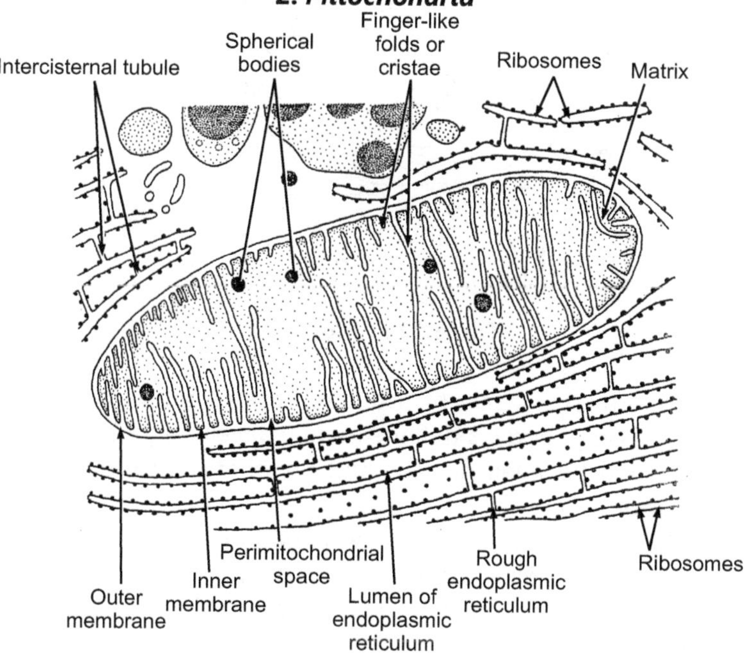

Fig. 4.21 : Endoplasmic reticulum and mitochondria

Comments :

1. Mitochondria are found in all eukaryotic cells.
2. It is sausage-shaped having outer and inner membranes. The inner membrane projects into matrix in the form of folds called cristae. The spacer between outer and inner mitochondrial membranes is called perimitochondrial space.
3. The inner membrane surrounds the inner chamber which is filled with matrix.
4. The cristae bear several F_1 particles.
5. The mitochondria are said to be originated from E.R.
6. The outer and inner mitochondrial membranes, outer and inner chambers bear several respiratory enzymes.
7. The mitochondria is called power house of the cell because they release energy by the oxidation of food stuffs and store it in the form of ATP.
8. The name mitochondria was given by Benda (1897-98) to the bioblast of Altmann (1890).

3. *Golgi Body*

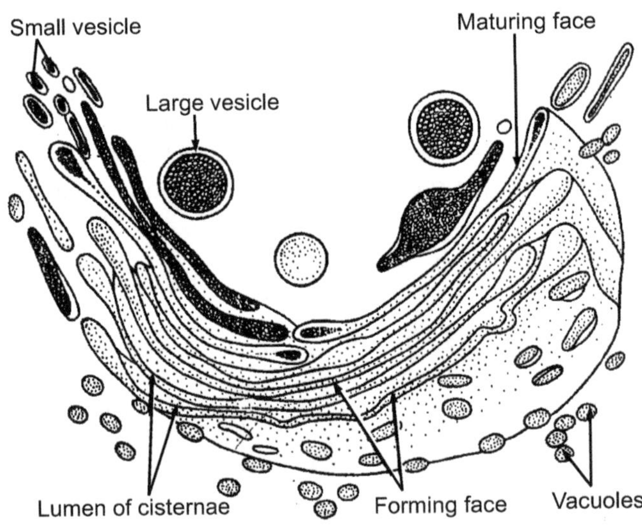

Fig. 4.22 : Golgi body

Comments :

1. Golgi body is formed of stacks of cisternae, vesicles and vacuoles.
2. The position of Golgi body is definite in animal cells, e.g., in the cells of ectodermal origin, it is found between nucleus and periphery, while in nerve cell it occupies circumnuclear position.
3. The secretory cells possess much active Golgi body.
4. It is very rich in phospholipids, proteins and enzymes.
5. It stores and concentrates various cellular products like proenzymes, carbohydrates, yolk, hormones, etc.
6. Lysosomes are believed to be formed from Golgi body.

7. It gives rise acrosome of sperm during the process of spermiogenesis.
8. It was reported by Camillo Golgi (1891).

4. *Nucleus*

Comments :

1. Nucleus (Fig. 4.23) is found in all eukaryotic cells except mammalian R.B.Cs.
2. It is usually spherical and formed of two membranes which are separated by a perinuclear space.
3. Both the membranes are fused together at a number of places so as to form nuclear pores through which nuclear and cytoplasmic contents remain in communication to each other.
4. It contains nucleoplasm, chromatin net and nucleolus.
5. The nucleolus is a spherical body containing RNA and nucleoprotein.
6. It controls all the vital activities of the cell and carries hereditary material from generation to generation.
7. It was first described by Robert Brown (1831).

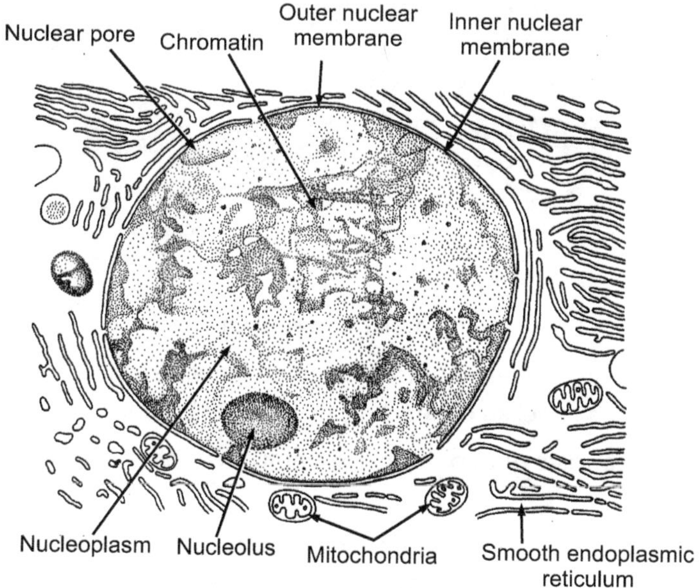

Fig. 4.23 : Nucleus

5. *Chloroplast*

Comments :

1. Chloroplasts are found in the cells of green plants.
2. It is a kind of chlorophyll-bearing plastid.
3. These are spheroidal or ovoidal in the leaves of higher plants, but may be club-shaped, stellate, collar-like or like a spiral band in algae.
4. On an average the chloroplasts in higher plants measure 4-8 microns in diameter.

5. In the cells of higher plants about 20-40 chloroplasts are found in each cell, but in algae usually one chloroplast is found in each cell.

6. Each chloroplast is covered by a double unit membrane. Each unit membrane is 40-60 Å in thickness.

7. The space between the two unit membranes is 200-300 Å.

8. This space is called periplastidial space.

9. The inner matrix of chloroplast is known as stroma.

10. Several chlorophyll bearing double membrane lamellae are found scattered throughout the length of chloroplast.

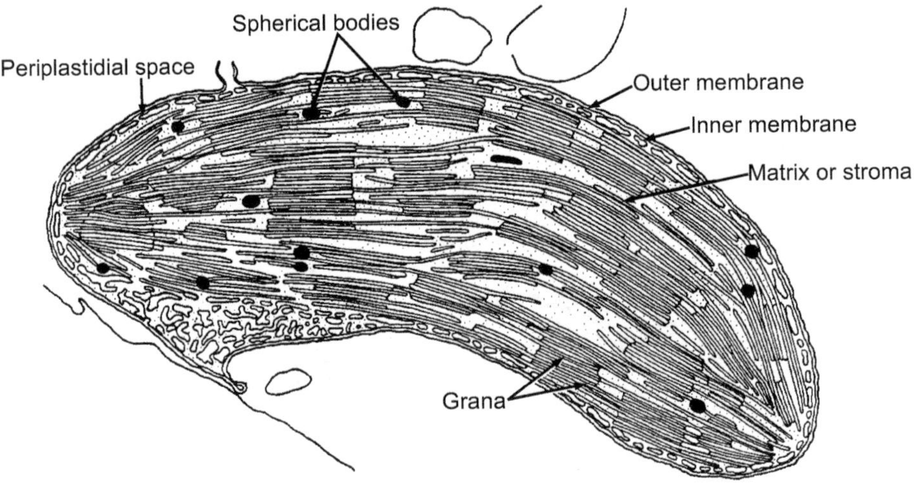

Fig. 4.24 : Chloroplast

11. Lamellae are of two types : (i) disc-like grana lamellae which are piled up at several places, (ii) stroma lamellae connecting the adjacent grana.

12. Each granum consists of about fifty superimposed membranous compartments, the thylakoids.

13. Chloroplasts carry out photosynthesis by trapping solar energy. These are said to contain enzymes which are required during Kreb's cycle and during synthesis of fatty acids.

✱✱✱